The Lighting of Buildings

THE
LIGHTING OF
BUILDINGS

R. G. HOPKINSON

and

J. D. KAY

FREDERICK A. PRAEGER, Publishers
New York · Washington

BOOKS THAT MATTER

Published in the United States of America in 1969
by Frederick A. Praeger, Inc., Publishers
111 Fourth Avenue, New York, N.Y. 10003

Library of Congress Catalog Card Number: 69-19889

Printed in Great Britain

Contents

List of Illustrations

List of Illustrations

FIGURES

Preface

The main purpose of this book is to discuss the work on lighting which was undertaken at the Building Research Station in Great Britain jointly with the architectural development groups at the Department of Education and Science and the Nuffield Foundation and later at the Ministry of Health. It is an attempt to bridge the gap between building research and practice.

The authors would like their readers to feel that they are participating in a seminar rather than submitting to a lecture, and that this introduction to the subject will arouse their interest and also go some way to complete their education. Principles are discussed, and some examples of solutions are given, but it is hoped that by being alerted to the problems and the means of solving them, the reader will have the interest and the greater satisfaction of also devising his own solutions.

There is very little about the hardware of lighting to be found in the book. Such information dates quickly and is best obtained from the literature issued by the manufacturers.

While the book was being prepared, decisions were made for Britain and her associates and colleagues in the Commonwealth to adopt the metric system. Both British and metric units are therefore used throughout the book.

Acknowledgements are gratefully made here to colleagues and others whose work has been drawn upon when preparing the text, but as the book is not intended primarily for research workers, it has been felt preferable for the continuity of the narrative to avoid too many detailed references in the text. The bibliography should assist reading in greater depth. The authors would in particular like to express their gratitude for the support and encouragement given to them by Sir Frederic Lea when Director of Building Research, and the late Anthony Pott when Chief Architect of the Ministry of Education.

R. G. HOPKINSON
J. D. KAY

October 1967

Introduction

The new generations of architect-trained environmental designers and building-orientated lighting engineers need a new kind of book in addition to the working textbooks and reference books of lighting technology. Good design is based not only on measurement and calculation but also on experience, observation, and intuition. The frontiers of lighting science have been extended into fields which were quite recently considered the province of the lighting artist, while in its turn the art of lighting has developed with the advent of new sources of light and new building and lighting materials.

Lighting was once one of the main modes of expression of the architect, and by lighting was meant exclusively day lighting. Robert Grosseteste, who inspired the design of Lincoln Cathedral, called light 'the most noble of natural phenomena, the least material, the closest approximation to pure form'. Even as late as the present century many schools of architecture limited the study of lighting to daylight and treated it entirely in artistic terms. The concern was with appearance and never with quantity. The quantitative aspects were left to the illuminating engineers, who emerged as a profession at the beginning of the present century, and with a few notable exceptions concerned themselves entirely with electric lighting. It is only recently that architects and engineers have come together in a common study of daylight and artificial light with equal emphasis upon the qualitative and the quantitative.

The origins of this book lie in the work which teams of architects, physicists, and engineers at the Building Research Station, and the architects of a few pioneer development groups under-took together during the years which followed the Second World War. The extension of technology out of the boundaries of classical physics and engineering into the field of psychophysics evoked a ready response, first in the group of young architects working at the Hertfordshire County Council who were pioneering new methods of school building designed to make learning attractive to the child. The collaboration developed to the field of hospital design where a group of architects at the Nuffield Foundation were breaking away from the traditional Florence Nightingale ward with its long row of beds and high windows. These two groups were later followed by the larger organisations at the (then) Ministry of Education and the Ministry of Health and subsequently elsewhere. The cross-fertilisation which resulted from this joint effort helped to advance the research programme and made architects appreciative not only of the usefulness of technology but also of its indispensability.

While it is essential to define the scope of the book and the philosophy behind it in this way, it is not easy to translate intentions into writing. Part of the difficulty is that different readers will have diverse technical background knowledge. Architects will not want to be forced to assimilate the detailed terminology of the lighting engineer, yet some of the concepts which were developed are highly sophisticated, and it is difficult to explain them fully and adequately without assuming knowledge of the fundamentals of vision and of the physics of light. The architects who collaborated in the work picked up the necessary technical information as the work proceeded as a child learns its mother tongue, but the architect new to the subject is in the position of having to learn a foreign language in a limited time. The technologists in turn learnt to think architecturally and above all to reconcile themselves to the limited role which lighting necessarily plays in total design of a building, perhaps the most difficult lesson of all. Lighting engineers who read the book and deprecate the apparently scanty treatment of matters which they consider of prime importance in their discipline will perhaps be tolerant.

The greatest problem has been to avoid the unnecessary repetition of what has been better said elsewhere, while at the same time making the treatment sufficiently comprehensive. The university student accepts the need to study with three or four books in front of him at a time, to which he makes constant cross reference, but the busy professional man can rarely find the time for this kind of intense study. Topics which demand more extensive reading for complete understanding have, however, been treated in sufficient detail for the main argument to be followed, even if reference must be made elsewhere if ideas are to be carried to complete design solutions.

1
Lighting and Building Design

LIGHTING AND THE BUILDING AS A WHOLE—'COMMODITY, FIRMNESS AND DELIGHT'

Sir Henry Wotton concisely stated the architect's problem over 300 years ago: 'Well building hath three conditions: commodity, firmness, and delight'. What is true for the whole applies also to the parts, and the lighting designer may find it useful to start from this standpoint today. The lighting of a building must clearly have 'commodity', that is, it must meet the social and physical requirements of the users. It must also have 'firmness': the technology must be sound. Beyond this, and perhaps most of all, the lighting must evoke 'delight', and play its part in inducing the desired emotions and creating an appropriate character for the building. All parts of a well-made building contribute to the whole, and the designer of one element—especially an all-pervasive one such as lighting—should appreciate the way it interacts with others.

Thus for the architect lighting is only one element in the design of a building, although a most important one. In a large building, the extent to which the general working illumination will depend upon daylight and how much upon artificial lighting will have a decisive effect upon the layout and planning of the building as a whole. The size and position of the windows and the type of glazing used will affect solar heat gain in summer and thermal losses in winter, and thus the heating and ventilation requirements. Likewise a multi-storey building which is deep in section and relies to a large extent upon artificial lighting will also have

23

to depend upon artificial rather than natural ventilation. It is therefore necessary to consider carefully the ways in which alternative methods of lighting may affect both the capital and running costs of a building.

DESIGNING FOR THE SENSES

It is desirable that the lighting designer should have an understanding of at least the main features of the complex physiological process by which we perceive our visual surroundings, and an account of this is given in Chapter 3. The architect should also consider in what ways there may be visual associations with other aspects of the physical environment that we perceive with our senses. Our sensations at the appearance of a textured surface derive partly from our acquired tactile experience. Certain colours are commonly associated with feelings of coolness or of warmth. The view of moving clouds and foliage through a large window is linked with the sense of freshness that can be given by natural ventilation through an open window. In a similar way our senses of hearing, taste, and even of smell can have visual associations that the lighting designer may need to take into account. In our studies of the physical environment we are now beginning to appreciate the implications for design of the fact that our subjective reactions to our surroundings are not always linked to a reading on a meter.

LIGHTING THE BUILDING

Light is a primary element in architectural design. Solid volumes and enclosed spaces, colour and texture can only be appreciated fully when they are imaginatively lit, and a skilful designer will use lighting in subtle ways to reveal the building. The most successful buildings are those in which the lighting of the building itself and the lighting of the activities it contains together make up a unified design concept, both by day and by night.

Lighting design depends largely on the control of brightness contrasts: the interaction of light and dark, texture (which is a pattern of highlight and shadow), and colour. The architect will normally use light and colour to clarify the form of building, to underline its structural logic, to help the user to find his way about, to draw attention to particular features of interest, and to point out hazards. And although lighting will usually be employed to reveal the building, the designer may also use it to conceal or to surprise. Lighting can be used to modify spaces in many different ways. It can reveal or conceal surfaces, heighten or diminish spaces. The theatrical lighting designer and the artist have often exploited lighting in this way, and there is much that the architect and the lighting engineer can learn from them (Plates 1 to 8). Picasso sees light as a 'measuring instrument in a world of shapes'.

VARIETY

The means for creating this diverse pattern of brightness—in addition to setting out the main geometry of the interior—which are at the disposal of the designer include:

  (i) varying the intensity of lighting from place to place;
 (ii) the choice of materials with different capacities for reflecting or transmitting light (whether they are light or dark, matt or gloss, translucent or opaque);
(iii) the position of the light sources and the distribution (direction of the flow) of light from them in relation to the main activities, surface, and objects in the room;
(iv) the colour both of the surfaces in the room and of the light emitted by the light sources.

The range of brightness in an interior created by these means can be considerable—the sparkle of cut glass or metal light fittings may be 10,000 times brighter than a black surface in the same room—and much of the vitality and interest of a lighting scheme will depend on this. At the same time, excessive contrasts in the field of view can cause visual discomfort, or actually reduce one's ability to see, in ways that will be explained in Chapter 3.

However, provided the mechanism of this is understood, lighting schemes can be designed in which the stimulus of variety is enjoyed, and at the same time the disadvantages of excessive contrasts are avoided.

LIGHTNESS AND DARKNESS

The American architect Leslie Larson, swimming bravely against the stream of high-powered illuminating engineering, observes wryly that the eradication of darkness has come to have moral overtones: 'The greater the intensity [of light] the more the assurance that darkness has really been banished and will not creep back round the edges.' In fact, darkness is the counterpart of light; each complements the other. A bright surface can only be appreciated fully when compared with a relatively darker surface, and vice versa. Simultaneous contrasts will appear to accentuate each other, as Leonardo da Vinci observed: '. . . if you see a white cloth side by side with a black one, it is certain that the part of the cloth which is next to the black will seem whiter by far than if the part is next to something whiter than itself. . . .'

Light became one of the central themes of J. M. W. Turner's later paintings, and he knew well how to sustain a suffused, luminous character—not necessarily of high brightness—by counterpointing light and dark (see Plates 3 and 4). Turner paved the way for the Impressionists and they too were able to create a sensation of brilliance in a medium whose range of actual brightness is very narrow compared with that available to the lighting designer. Gombrich draws attention to the importance of gradients, rather than objective brightness: 'Wherever we observe a sudden steep rise in the brightness of a tone we accept it as a token of light.' And when we see (as in Plate 1) an evident source of light in the windows, as well as a sharp gradient of brightness in a surface which our experience tells us is homogeneous, then we feel sure that the bright patches on the floor are caused by direct sunlight.

Turning from painting to building, an architect has the advantage that he can choose to brighten a wall either by making its surface reflect more light compared with its surroundings, or by flooding it with light—or even both together. Where it is absolutely necessary that an interior should, because of its mode of use, have an even level of illumination, then giving one surface a higher reflectance may be the only way of giving it preferential brightness. If on the other hand a surface—say a floor—has necessarily to be of a single, uniform material, then the only way to give parts of it preferential brightness is to increase the illumination locally. It is not, of course, necessary to reveal the source of the extra light; a brilliantly lit surface can have greater impact if the window or light fitting is concealed (and is not in danger of detracting from the object lit). However, if it is desired that we should perceive a surface as having a constant colour—for example, a wall running back from a window down the length of which the level of illumination will fall off—then the source of light should be seen to be there. This phenomenon of brightness constancy is discussed further in Chapter 3.

CHOICE OF MATERIALS

From what has already been said, it is obvious that the choice of materials with different reflectances can have a marked effect upon the pattern of brightness in an interior. This not only affects the brightness of each surface but also the proportion of indirect (i.e. reflected) light in the room. For example, in a schoolroom daylit from one side only, and with light interior surfaces, as much as half of the illumination at the back of the room can be from inter-reflected light. Not only is the level of illumination affected, but also the appearance of objects in the room. A room with a high average reflectance and consequently a high proportion of indirect light will have soft contrasts and gentle modelling. On the other hand, an interior with dark surfaces will have a high proportion of direct light and a harsh, dramatic character with strong contrasts of light and shade. Each approach to lighting design has its place.

A common problem in lighting is that caused by the reflected images of bright light fittings or windows in smooth, shiny surfaces such as gloss paint, lacquered wood, plastics, or highly polished floors. This reflected glare will be most troublesome on shiny surfaces that are dark and such materials should not be used for desks and other working surfaces, and should be avoided on floors. The reflection of light fittings in uncurtained windows at night can also be irritating. Under normal circumstances the main surfaces of a room—walls, floor, and ceiling—should have a matt finish in order to avoid disturbing reflections.

In most working interiors polished materials are best confined to small-scale items—metallic fittings, door furniture, and glass-painted architectural 'trim' such as architraves, skirtings, and window frames. The sparkle and highlights will make for a lively interior scene.

The materials discussed so far are opaque and reflect light. Materials that transmit a proportion of light are used chiefly to shield artificial light sources, most of which are too bright to be viewed directly in comfort, or to screen windows. In the design of light fittings in addition to the 'traditional' materials of wood, metal and glass, a wide range of plastics is now available with various degrees of opacity and in extruded or pressed mouldings. These are now available in prismatic forms that make it possible to deflect light in desired directions, and to reduce brightness from normal angles of view. Similarly, an increasing range of materials for window blinds is now available, in both natural and artificial woven materials and in slatted and 'venetian' forms. It is now possible to specify the amount of light to be transmitted and the brightness of the blind when lit. The design aims for materials used in light fittings is discussed further in Chapter 4 and those for window blinds in Chapter 5.

DIRECTIONAL LIGHTING

The degree to which lighting helps to reveal the building depends largely on its directional qualities. Light which 'flows' through a building and across surfaces will model solid forms and accentuate textures. Directional lighting can heighten our appreciation of space and materials, while lighting which is too diffuse will diminish it. In addition to helping to articulate the building, directional lighting will help to place objects and to clarify their shape and details. Directional lighting may perform many roles in a building and will take various forms, even in one room. Specifying and designing for directional lighting is thus a complex process. In each case the designer has to decide which should be the main direction of 'flow' of the light, and how 'strong' it should be (that is, the proportion of direct to indirect lighting).

Recently attention has again been drawn to the fact that the engineering convention of measurement of illumination on a horizontal plane rarely correlates with one's subjective assessment of the adequacy of illumination in an interior. This assessment depends more upon the apparent brightness of things seen around one. It relates more closely to the quantity of light arriving at the given point on surfaces facing all directions, rather than that measured on a horizontal surface only. This 'spherical' illumination is called the *scalar* illumination. However, it gives no indication of the directional characteristics of the lighting, and it is necessary to supplement it by a directional measure 'the illumination *vector*' in order to give a better quantitative description of the assessment by the eye. The directional strength of the lighting would then be indicated by the ratio of the illumination vector to the scalar illumination.

These measures go a considerable way towards making it possible to specify a given direction and strength of lighting, with an understanding of its likely effect on the appearance of the interior.

There is, however, much more work to be done in order to correlate quantity and quality. There are, for example, differences in modelling and shadowing which result from a single large source, compared with, say, four smaller sources distributed over the same area. Both alternatives could have the same scalar and vector values but the former would tend to be 'shadowless', while the latter would produce multiple shadows.

The majority of people find themselves more satisfied with the appearance of an interior when the light flows mainly from one direction, and most people prefer that it should come neither horizontally nor vertically downward, but diagonally—over one shoulder, as it were. This seems to be preferred because it models the human face in an agreeable way, and because it makes it reasonably easy to avoid working in one's own light. This diagonal flow, is in fact, very much that given by normal windows, and it is very difficult to simulate this in an acceptable manner over a whole room by artificial lighting. Regular arrays of lighting fittings over a large room will produce a resultant flow of light that is vertically downwards. This may be acceptable as a second best provided there is a large indirect component to fill in the heavy shadows that would otherwise be produced by downward lighting that is too strongly directional. For this reason the average reflectance of working interiors should be kept high.

In deep rooms where daylight needs to be supplemented by permanent artificial lighting (see Chapter 6), the windows can provide this valuable inclined directional component, in addition to allowing a view out, and they should be designed to do this.

Local directional lighting can usually be provided most easily by suitable artificial light fittings, whether the aim is, say, to light a chalkboard, to draw attention to an information desk, or to dramatise a piece of sculpture.

COLOUR

Too many lighting schemes—and for that matter too many buildings—still seem to be designed initially in black-and-white, a colour scheme being superimposed later. The capacity of a surface to reflect light ('Value' in Munsell terminology), may perhaps be considered when designing the lighting, but its other colour attributes (Munsell 'Hue' and 'Chroma') are not.

An architect will normally use colour to underline the structural logic of the building and to clarify the way in which it is to be used, as well as to create a varied and interesting interior with the appropriate character. Increased colourfulness can also be used to draw attention to particular features, in the same way that increased brightness can. It is obviously desirable that the lighting of a building should not contradict the colouring of the building, but rather that it should enhance it. To ensure this lighting and colouring should be designed together from the start.

For example, where subtle differences in colour have to be appreciated, either in the work that is being carried out or in the decoration of the building, a higher level of illumination will be required. The colour of the artificial light source will have to be chosen carefully where accurate colour rendering is important, or where work involving colour discrimination is being started by daylight and then carried on under artificial lighting, or where daylight and artificial light are used together in a room. Where for good reasons variety cannot be provided by the lighting in the ways that have been discussed above, then contrasts in colour can help to give the necessary stimulus and interest. Small areas of a dark, rich colour can be a bass note, as it were, to counterpoint an otherwise high key colour scheme.

LEARNING FROM EXPERIENCE

Much can be learnt from a study of old buildings provided we understand the circumstances in which they were built and the purpose for which they were originally used. Nearly a century ago Owen Jones said: 'The principles discoverable in the work of the past belong to us; not so the results. It is taking the ends for the means.'

Plate 2. The use of a single light source as a focal point has often fascinated painters. Joseph Wright of Derby was in touch with the upsurge of science in the second half of the 18th century, and in his painting of 'The Orrery' (this mezzotint was by William Pether, 1768), the use of one lamp to represent the sun in this astronomical experiment has been exploited both in the wide-range modelling of the figures (from the burnt out full face of the little girl to the figure in silhouette opposite her), and also in the hypnotic gaze of the watchers at the source hidden from us. (By courtesy of the British Museum.)

In Turner's later years his main interest became the transitory effects of the natural elements—air, fire, water—and in particular in light and colour. These informal sketches opposite were done in Petworth House about 1830.

Plate 3. The Morning Room is suffused with soft light with a high proportion of indirect illumination enlivened by the highlights of white paint on door frames and book-cases, and by the sparkle of crystal chandelier, top right.

Plate 4. Firelight and Lamplight is in a low key, with, on the left, the same device used in Plate 2 of a hidden lamp lighting or silhouetting the figures around it, and on the right—scarcely recognisable in the reproduction—a group of darker figures in the warm glow of a fire, the only other source of illumination in the room. (By courtesy of the British Museum.)

Plate 5. Picasso's use of light is of a different order from that in the preceding illustrations. In his 'Still Life by Candlelight' the observed facts of light and shade, and of highlights on the brass candlestick and gilt picture frame, have been the starting point, but Picasso has subordinated these to the design as a whole. Moreover, by giving the candle flame a black halo and by reversing the expected direction of the shadows on the cup and saucer, Picasso has heightened our awareness of the nature of light. (By courtesy of Galerie Louise Leiris; copyright by S.P.A.D.E.M., Paris.)

Plate 6 (opposite). 'Supernovae' by Victor Vasarely. This static design upsets our perceptual system sufficiently to create an illusion of pulsating movement which is most apt to its title. The reason for this is discussed by R. L. Gregory (see Bibliography). Recently artists have again become interested both in the deliberate exploitation of optical illusions to create ambiguity and a sense of movement, and in the use of light itself in motion as a part of their medium. (Reproduced by permission of the artist. Photography by courtesy of the Tate Gallery, London.)

Plate 7. 'Perpetual Mobile' by Julio le Parc. Spotlights are trained on polished aluminium plates hung on nylon wires which move at random in natural air currents. The patterns of light created are—within the geometry of the design—accidental, but certain similar patterns tend to recur and make the experience of looking at it a coherent one. (Photograph by courtesy of the Tate Gallery, London.)

Plate 8. 'Box D.11' by John Healey. Continuously changing patterns of light are created by discs revolving slowly in front of projected slides, in this case on a seven-hour cycle, so that no repetition is apparent. This is clearly only a short step from cinema, and we are again—as in the Baroque age—at the point where the art of the painter and sculptor meets that of the theatrical designer, and in ways very relevant to the architect and lighting engineer. (Photograph by Sloman and Pettitt.)

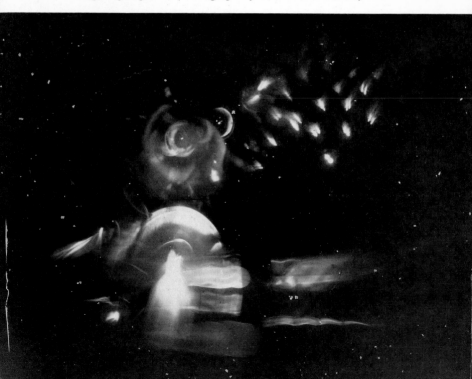

Developments in window technology up to the present have been all in the direction of larger and larger glazed areas. The displays of technical virtuosity in the design of very large windows which were characteristic of the first half of this century are now being tempered by the realisation of the environmental problems created by such large areas of glazing. The first of these environmental problems to be recognised was the loss of internal heat through a single skin of glass. More recently increasing traffic noise has aggravated the problem. As a result, the designer has had to cope simultaneously with complaints of excessive noise, sky glare, excessive solar heat penetration, loss of internal heat, and inadequate ventilation. There are many suggested solutions to this problem, but the complete solution must depend upon all aspects of the environment taken together and not only on the lighting. The complexity of the window problem in the modern world should not lead to a disregard of the benefits of a visual link with the outside world.

ARTIFICIAL LIGHTING

Artificial lighting during the greater part of our architectural history had little or no effect upon the design of the building. Until recently natural lighting was the predominating influence. When it was dark, the majority of the population just went to bed. The cost and the technical limitations of the available light sources such as oil lamps and candles was too great for them to be of any general application except in the houses of the rich. Gilbert White, in the *Natural History of Selborne*, describes how to make rush candles so that the labouring classes could have artificial light at a cost which was a not unreasonable proportion of the weekly wage, but this would be possible only in the country where natural materials were readily available. In the eighteenth century the invention of the glass lamp chimney was a major step forward, but it was not until the nineteenth century with the coming of town gas and the invention of the gas mantle that artificial lighting could be considered as a means of prolonging the length of the working day, or of extending leisure pursuits after dark.

If artificial lighting had no effect on the design of buildings, nevertheless the design of the light fittings themselves often reached a high level—brass, gilt, and porcelain-enamelled candelabra, elegant in themselves and exploiting the sparkle and flicker of the innumerable small point sources set a standard for which we are hard put to find a modern equal. But one must also remember that such a method of lighting had its disadvantages: soot-darkened ceilings, splashes of wax, the smell and the heat, and above all, the very high cost.

It took a long time for the design of electric light fittings to throw off the traditional forms carried forward from the candle and gas eras, and even longer for designers to appreciate the value of integrating much of the lighting with the fabric of the building, so that the result is just light, rather than a building cluttered with light fittings.

NEW DEMANDS ON LIGHTING

Since the days in which Sir Henry Wotton lived, social and economic developments have called for new types of building such as schools, hospitals, and factories in which the most significant difference from the point of view of their lighting is that they include large rooms and space where many people need to carry out similar visual tasks simultaneously. The spinner no longer sets her loom in the doorway of her cottage. She works with many other spinners in a mill where the lighting that she has depends on her position in relation to the windows or to the artificial light fixtures in the ceiling. The monk once took his manuscript to the window of his cell, but the modern clerk works in a large office with fifty more doing the same work and he has to put up with the lighting provided for him at his work place. At one time the craftsman could move to the light, now we must bring light to him at his place of work. Freedom of choice has gone, but by a careful study of what people really need, the right lighting can be provided for each and all of them at their place of work.

Since Wotton's day, technical developments in methods of building construction, glazing techniques, and more recently in electric lighting, have provided the means to meet these requirements for lighting large rooms by day and by night. The relationship between good lighting and efficiency in work has been understood and acted upon. The human eye and its functions remain the same, but optical aids such as spectacles are more efficient and our understanding of visual disabilities has improved, so that we are now better able to provide the best working conditions for the most difficult visual tasks and to extend effective vision into old age so that we can make use of valuable experience. An understanding of the nature of human vision and perception is the starting point of good lighting design.

2

Some Simple Lighting Concepts

RADIATION AND LIGHT

The phenomenon we call light is that very small part of the whole spectrum of radiant energy which exclusively has an action on the eye that results in vision. Radiant energy of different wavelengths gives rise to the sensation of different colours, progressing with increasing wavelength from violet, blue, and green, through yellow, orange, to red, and thence to the infra-red heating radiation which results in no sensation of light. The human eye is most sensitive to radiation which causes a yellow-green sensation (Fig. 1). This is one reason why artificial lighting of 'high efficiency' has a greenish appearance (early types of fluorescent lighting) or yellowish appearance (sodium street lighting).

Until quite recently light resulted from some form of incandescence, that is, radiation from a body at a high temperature. The sun is, of course, such an incandescent source, and so are artificial sources which give out light as a result of combustion (candles, oil lamps), or by the electrical heating to high temperature of a tungsten filament wire (the ordinary light bulb). Such incandescent light sources emit radiant energy throughout the spectrum. A mixture of all spectral colours in approximately the same proportions appears subjectively as white. Incandescent sources appear white with a bias towards the red and orange if the temperature is relatively low, and towards the blue if the temperature is high. An incandescent source progresses through 'red heat' and 'yellow heat' to 'white heat' as the temperature is increased.

Fluorescent lamps emit light by an entirely different process.

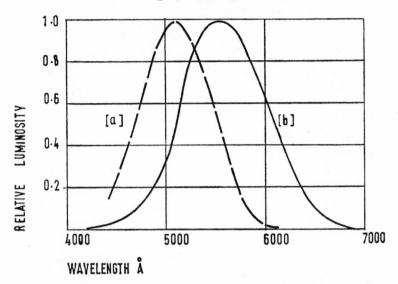

Fig. 1. The eye's response to different wave lengths of equal energy (after W. D. Wright). The eye responds only to the wave length of electromagnetic radiation covering one single octave, and has its maximum sensitivity under daylight conditions to light of a wave length which gives rise to a colour sensation of yellow-green (photopic curve (b)). It has a very low sensitivity to the violet and the red, beyond which limits lie the ultra-violet and the infra-red which affect the skin but do not create a sensation of light. This is the sensitivity of the 'cone' receptors of the retina, which operate only at high levels of light and in which resides the sensation of colour. By night, the cones which are relatively insensitive, serve very little to aid sight (there is no sensation of colour by moonlight) and the 'rod' receptors then come into action. The sensitivity of the rods is greater in the blue region of the spectrum (but they do not 'see' blue, only light) and less in the red compared with the cones (scotopic curve (a)). In the late dusk, red geraniums look black and deep blue delphiniums look almost white.

A sealed glass tube containing mercury vapour is provided with an electrode at either end, and on the application of electrical energy at the electrodes, the atoms of the mercury vapour filling undergo certain disturbances which result in energy being discharged which is converted into visible light by the fluorescent coating on the inner surface of the tube. The light emitted by the fluorescent tube depends upon the nature of this coating, and by suitable choice of materials, almost any colour from saturated individual hues to white or near-white colours matching daylight or incandescent light can be obtained. The skill of the manufacturer of fluorescent lamps is to obtain a colour which is suitable for any particular purpose such as, for example, providing a permanent supplement to daylighting to be used in a building during daylight hours so that the appearance of the room is as if natural daylight was the sole source of illumination. Fig. 14. compares the spectral power distribution and colour rendering properties of different artificial light sources.

The ideal lamp would be one which emitted radiation only in the band of wavelengths which give rise to a visual response. In practice all lamps emit radiation in the unwanted ultra-violet and infra-red regions which excite no visible sensation and so a factor of great importance is the 'efficacy' of a lamp in terms of the proportion of the total radiation which is emitted in the form of light. Incandescent filament lamps have a much lower efficacy than fluorescent lamps because a large proportion of their radiation is in the form of heat. Consequently fluorescent lamps give more light for a given expenditure of electrical power.

The unit of light is called the *lumen*. (The lumen is defined in terms of an absolute standard.) The light output from any lamp, or from any lighting fitting, is expressed in lumens. For example, a simple wax candle gives out about 13 lumens, a 100 watt filament bulb emits about 1200 lumens while a high efficiency 80 watt fluorescent tube emits about 4500 lumens.

ILLUMINATION

The lighting level, that is, the amount of light spread over the area of the work, is called the *illumination*, and is expressed in *lumens per unit area*. One lumen per square foot is the amount of illumination given by one ordinary wax candle placed one foot from the work. Unobstructed daylight from a bright overcast sky gives about 1000 lumens per square foot. The illumination on a drawing board in a well-lit office drawing will be not less than 50 lumens per square foot. The unit of illumination in metric countries is the lux, that is, one lumen per square metre. 1 lm/ft² is equivalent to 10·76 lux or, very approximately, to 10 lux.

BRIGHTNESS

Light falling on a surface is reflected back according to the ability of the surface to reflect light, called the *reflectance*. A white surface has a reflectance of nearly 100%, while a black surface has a reflectance of only about 2%. A medium grey has a reflectance of about 40%. If the surface diffuses light equally in all directions (for example matt white distemper) the brightness of the surface will be directly related to the illumination falling upon it. The relationship is:

$$\text{Brightness} = \text{Illumination} \times \text{Reflectance}.$$

The physicist and engineer use the word 'luminance' to express the physical measure of brightness. The unit is the foot-Lambert, and the relationship above then becomes:

$$\text{Luminance (ft-L)} = \text{Illumination (lm/ft}^2) \times \text{Reflectance}$$

Thus a diffusing white surface of 50% reflectance receiving 10 lumens per square foot will have a brightness of 5 ft-L.

Illumination to a level of 100 lm/ft² falling on a white surface of 80% reflectance, a blue surface of 30% reflectance, and a dark purple surface of 10% reflectance would have respectively luminances of 80 ft-L, 30 ft-L, and 10 ft-L.

With the more widespread adoption of the metric system, an internationally agreed unit will have to be adopted for luminance relative to the illumination (lux) on a surface. This 'equivalent lux' is called an 'apostilb' (asb) in central Europe. The suggestion has been made to use the term metre-Lambert (m-L). The former will be used here. Thus:

$$\text{Luminance (asb)} = \text{Illumination (lux)} \times \text{Reflectance}$$

COLOUR

Reflectance is a measure of the *total* light reflecting properties, and so surfaces of widely different colour can have the same reflectance.

The colour properties of a surface can usefully be expressed in terms of their Munsell co-ordinates, the Hue (red, green, blue, etc.) and the Chroma (that is, the colourfulness). On the Munsell system lightness or reflectance is expressed in terms of a quantity called Value, related to reflectance by the formula $R = V(V-1)$, so that a colour of Munsell Value 6 has a reflectance of $6(6-1) = 30\%$. Fig. 2 shows the relationship of the Munsell Hue, Chroma, and Value scales.

Due to the fact that the eye adapts to its surroundings, and because its response is greatly influenced by adjacent contrasts, the subjective effect of light is not directly linked to its physical amount. A piece of grey card placed on a white surface looks much darker than when the same grey card is placed on a black surface. More generally, the *apparent brightness* of any object in a scene depends not only upon its physical luminance but also upon the surroundings in which it is seen. A car headlamp looks intensely bright in the dark surroundings of an unlit street at night, whereas the same headlamp seen in the bright sunlight by day appears as of nothing like the same high brightness. Consequently in designing lighting it is necessary to distinguish between the *physical brightness*, or luminance, which is measured by a physical instrument and is expressed in physical units, and the *apparent brightness*, which is the actual subjective effect that we see, which has to be expressed in arbitrary subjective units which are related not only to the physical luminance of the surface itself, but also to the physical luminance of the whole scene which governs the adaptation level of the eye. Lighting has to be designed for what we see.

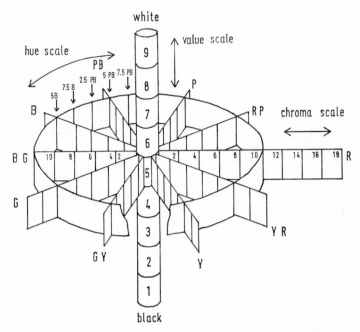

Fig. 2. The Munsell system of colour specification. The Munsell system of colour specification enables any colour to be identified in terms of the three main attributes of colour that are significant for a designer:

—its *Hue*, that is, whether it is red or green, etc.
—its *Value*, that is whether it is light or dark*
—its *Chroma*, that is, whether it is a strong or weak colour.

For example, a strong 'poppy' red might be identified thus:

7·5 R	5	/	16
Reddish hue	Medium value		High chroma
tending towards	R = 20%		
yellow–red			

This system is used for referencing the colours in B.S. 2660. Munsell colour atlases are available.

*Munsell Value is linked to reflectance by the formula:

$$R = V(V—1)$$

Where R is the reflectance % to white light and V is the Munsell Value.

3

Lighting and Vision

A room appears pleasing to us because the play of light and colour evokes satisfying sensations. On the other hand, the ability to read the small print of the newspaper depends more than anything else upon the actual amount of light falling upon the print. Consequently in designing for appearance, we are concerned essentially with brightness and colour, while in designing for sharpness of vision we are concerned with amounts of light. This distinction is at the root of all lighting technology. It is a useful distinction because it enables many simplifications to be made in the design processes.

SIGHT AND INTELLIGENCE

Textbooks of physics usually compare the eye to a camera. The eyeball itself, with its lens and its sensitive film, the retina, is certainly a camera, but it is very much more. What we see depends not only upon the image which is focused upon the retina, but on the intelligence which interprets that image. Nowadays there is the electronic computer to use as an analogy for the visual sensory processes which perform the interpretation of the retinal image. The analogy is a good one, except that it leaves us with an overwhelming sense of awe at the unbelievable complexity of any computer which would even begin to match the functions of the simplest vertebrate eye, let alone the much more complicated seeing process of the human being.

The comprehension of the two drawings shown in Figs. 3 and 4 does not rest with the eye alone. Fig. 3 consists entirely of

Fig. 3. Perception and experience: I. Whether the whole story of this picture is seen immediately depends not so much on sight as on experience and interpretation. The black blobs can be seen clearly even by a poor-sighted person, but only familiarity with pictures of medieval plumed knights on horseback permits the full interpretation. The illustration is derived from R. S. Woodworth Experimental Psychology (New York 1938).

Fig. 4. Perception and experience: II. In this picture it is not because of missing information (as with Fig. 3) that the full interpretation is difficult, but because there is too much. The title 'My Wife and My Mother-in-law' by W. E. Hill, supplies the necessary clues, and the intelligence behind the eye then makes the interpretation. Whether one's wife or her mother is seen first probably depends upon one's present preoccupations.

patches of black on the white background, which can be seen without difficulty. On the other hand, the interpretation of these patches depends on the ability to fill in the missing detail, and this ability differs from person to person. Some people see immediately the knight in armour on his plumed charger, while others cannot even after detailed description has been given. Equally the detailed drawing of Fig. 4 may reveal one thing to one person and another to another person as the equivocal title indicates. The picture conveyed by incomplete information such as Fig. 3, or super-abundant information as on Fig. 4, is not a matter of the visual function alone, but of the interpretation which is put upon the visual information by the intelligence behind the optical system of the eye.

In the same way, the interpretation which we make of an ordinary visual environment will depend not only upon our visual capacity for registering a sharp well-detailed image, but also on the intelligence which we bring to bear on the interpretation. It is known, for example, that someone who has been born blind and who has subsequently had full vision restored in later life does not see the environment in three dimensions as do the rest of us. He has to make a conscious effort to learn to see, and to correlate familiar tactile information with the unfamiliar visual information now being received for the first time. This effort may be very great, and in at least one well-documented case resulted in complete nervous breakdown and subsequent suicide. Those of us who have been gifted with the sense of sight all our lives consequently fail to realise what a tremendous amount of visual skill we have accumulated during this period. It is comparable, though on a far greater scale, of the effort required to become thoroughly fluent in a foreign language as compared with the ease with which we learn our mother tongue.

ADAPTATION

The dominating feature of human vision is adaptation. Everything we see is referred to some reference level whether of light, or darkness, or colour, and we make our interpretation in terms of this adapting reference level. All visual experience, whether of brightness, or of colour, or of distance, or of perspective, or of solidity, must necessarily be referred to some norm of past or present experience. The importance of this in the field of lighting

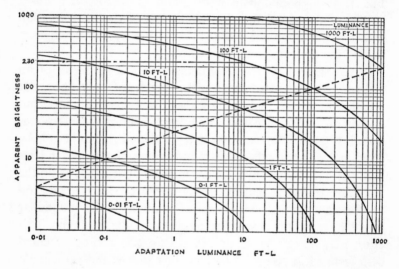

Fig. 5. Adaptation level and apparent brightness. In any given scene, the eye sensitivity settles down to a general average state of adaptation. This acts as a 'reference standard' such that individual items of the scene which have a higher physical luminance than this reference level 'look bright', and those with a lower luminance 'look dark'. The brilliance of the highlights and the murkiness of the shadows consequently depends not only on their intrinsic physical luminance, but also on the state of adaptation to the eye. Raise the adaptation and the shadows look darker. Lower the adaptation (screen the window with your hand) and the shadows look brighter. So do the highlights. Thus in the diagram a surface with a luminance of 100 ft-L has an apparent brightness of 100 when one's eye is adapted to 100 ft-L, but the same surface would have an apparent brightness of 230 when one's eye was adapted to 10 ft-L.

is immense. What we see depends not only upon the actual physical amount of light or of colour present, but also upon the state of our eyes at the time of seeing, and upon the amount of visual experience which we have to aid us in our judgments.

This problem of adaptation of the eye also comes into prominence in the design of daylighting. Whereas at night the artificial lighting is entirely under the control of the designer—he can put in a large number or a small number of lamps, of whatever power he chooses—during daytime the interior lighting is dependent only partly upon the designer's intentions. He can determine by the size and position of his windows, and by the lightness (reflectance) of the interior surfaces what *proportion* of the available daylight he will admit, but he cannot determine its absolute *amount* because this is governed by the variable light available from the sun and sky, i.e. with the time of day and year, the weather and the cloudiness, vital factors which are not under control. Fortunately these wide var ations in daylight do not result in such large changes in the subjective effect in the interior, because the eye can adapt very readily to slow changes of lighting level. In fact, on an overcast day the eye is almost unable to estimate how much light there is, a fact which apprentice photographers who are foolish enough to guess their exposure without the aid of a light meter soon learn to their cost. Interior daylight is therefore measured in terms of the Daylight Factor, that is, the ratio of the interior illumination on a given surface at a given reference point, to the total illumination available from the whole unobstructed sky outdoors. This Daylight Factor is therefore a measure of the proportion of total light which is available indoors, and not of the actual amount of light. It is a better measure, in fact, of the visual effect than would be the actual illumination provided the eye has a view of the sky outdoors through the window to determine its state of adaptation. As the sky brightens, the eye adapts to the added light, and as the sky darkens, the eye dark-adapts so that the visual effect is more nearly related to the proportion of available indoor light than to the actual amount.

The adaptation of the eye to light and colour is a highly complex physiological process. It is not merely the small effect which results from the opening and closing of the pupil, which is really nothing more than an 'early warning' system which helps to protect the eye from harm while other slower acting processes of the retina come into operation.

The lighting of interiors is concerned with only a small part of the available adaptive range of the eye. The whole compass of interior lighting covers a range of perhaps 10,000 to 1 in light quantity from the lighting level produced by unobstructed direct sunlight to the lighting of a shaded domestic interior and, given time, the eye can adapt itself to operate efficiently at either extreme. Even so, the eye cannot function at its best in bright sunlight and deep shadow simultaneously, that is, not all of the adaptive range of the eye is functioning at any one time. If one looks out of the window and adapts to the bright scene outside, the clouds, the sky, houses, and trees will all be seen clearly, but the darker shadows indoors will appear devoid of detail and the interior will look gloomy until the eye has properly re-adapted to the interior conditions. A gloomy appearance is much more likely to be due to unfavourable adaptation than to lack of light.

One of the essential requirements of good lighting is to avoid creating situations where the eye is called upon to adapt too quickly over too wide a range. This is especially true if any problems of safety are involved, but even if they are not, great discomfort can result if the eyes have to look at brilliant areas of sky and then immediately back at a book, a chalkboard or a machine. Unfavourable adaptation is one of the pitfalls of lighting design. The understanding of the process of adaptation is perhaps the most important principle that the designer has to grasp.

At any given instant when the eye is comfortably adapted to the whole scene, the range of brightness which is appreciable is limited at the lower end by the inability to perceive contrast in shadows, and at the upper end by a corresponding inability to perceive detail in excessively bright highlights due to blinding glare. The precise way in which the adaptation level acts as a

Plate 9. New forms of daylighting to large workrooms were called for when the Industrial Revolution concentrated weaving into large factories. This early 19th-century woollen mill at Nailsworth, near Stroud, Gloucestershire is now Bentley's Piano Factory. (Photograph by Eric de Maré.)

Plate 10. The Oxford Museum (1855) is a unique example of the application of the new iron-and-glass techniques (with a mock-Gothic veneer) to the display of advanced (Darwinist) science. The uncontrolled quantity of light has disastrous effects on some of the exhibits. (Photograph by Eric de Maré.)

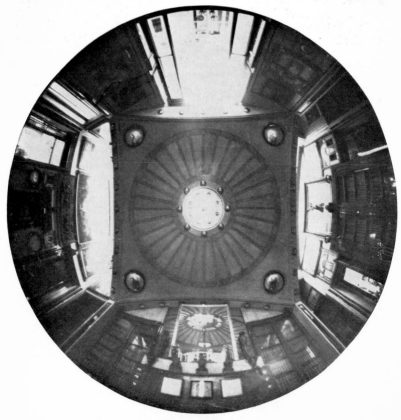

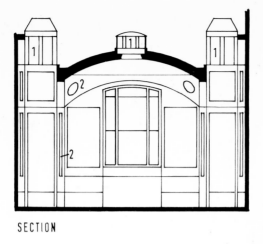

SECTION

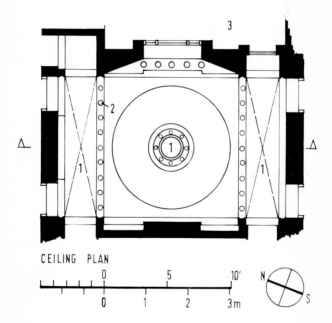

CEILING PLAN

Plates 11, 12 and 13. The daylighting in the Breakfast Room of Sir John
Soane's own house in Lincoln's Inn Fields (now the Soane Museum) is
original and subtle. The main source of light is a window to the right of the
photograph above left (Plate 11) and picks out the characteristic incised
decoration on the shallow saucer ceiling. Light from concealed roof
lanterns (1) also floods down the walls on either side. The feeling of space
is extended by the concealed boundaries to the room, by the use of mirrors
(2), and by distant views through glazed doors and into the courtyard (3).
(Plate 11 by courtesy of the Curator of the Soane Museum. Photograph by
Sidney W. Newbery. Plate 12 full-field photograph on Robin Hill camera
by J. Longmore.)

Plate 14. The architect of the new building for the Royal College of Surgeons was asked to provide a setting for the old, panelled Council Room. The strongly directional daylight enhances the appearance of the room, and the dark panelling sets off the picture frames and chandelier. The slanting light models the sculpture (and presumably the Council members) in an agreeable way. The whole is strikingly similar in character to Plate 1. (Architect: Denys Lasdun. Photograph by John Donat.)

Plate 15. The light flooding down on the altar of this church in Glenrothes from a concealed lantern tower forms a dramatic focal point for an otherwise austere interior. (Architects: Gillespie, Kidd & Coia. Photograph by Wm. J. Toomey.)

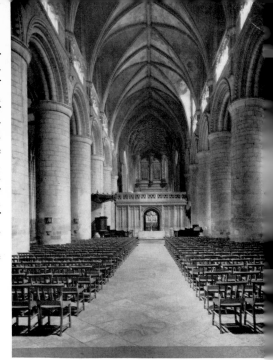

Plate 16. In the re-lighting of Gloucester Cathedral particular attention was paid to the lighting of the fabric of the building in addition to ample illumination on the floor of the nave. The lighting designer (J. M. Waldram) has chosen to follow the flow of light through the building from south to north which occurs during the day. This is achieved mainly with simple projector floods placed in the triforium. The modelling emphasises the solidity of the columns in the nave, and throws into relief the details of the screen. (Photograph by courtesy of the General Electric Company Ltd. Photograph by Fox Photos, Ltd.)

Plate 17. Special lighting techniques have been used to dramatise the appearance of the small Gethsemane Chapel in Coventry Cathedral. The strip of artificial lighting which separates the ceiling from the wall is on during the day and after dark, and gives both texture and a luminous depth to the mosaic mural. Daylight enters through the rooflight and between the mullions on the right. The architect (Basil Spence) and lighting designer (John Reid) have used contrasts of lighting and colour, of polished and rough materials to dazzle and awe the visitor. (Photograph by John Maltby, F.I.B.P.)

The eyes can also adapt to different colours and in particular to 'white' light of different forms. The light of a candle, or of a filament lamp, a fluorescent lamp, or overcast skylight or sunlight are all of a character which we would call 'white' because when the eyes are fully adapted to any one of these forms of white light *with none other present* white materials like paper appear white rather than coloured. Mankind evolved in daylight and so our eyes work most efficiently in the light of the quality given by natural daylight. The light of the earlier artificial sources like the oil lamp, candle, gas lamp, and electric filament lamp all give light which, although a poor approximation to daylight, nevertheless has the important characteristic that they are incandescent sources and show a continuous emission of all the spectral colours from red through yellow and green to blue and violet and so they compound to give the sensation of white. The eye can perform the act of 'colour adaptation' which permits colours seen in the incandescent artificial source to be judged with a fair degree of accuracy and related to the appearance of the same colours in daylight. It is true that some colours, particularly purple, are recognisably distorted in such a way that judgments are affected, but nevertheless the distortions which occur are generally acceptable.

Modern light sources such as the fluorescent tube, and other forms of discharge lamp, however, which do not depend on the phenomenon of incandescence, have brought in their train the need to think more precisely about colour and the rendering of surface colours of objects by reflection. These new light sources have become widely used because they are capable of giving very much more light, perhaps five times as much or more, for the same cost, than was possible with the best of the older incandescent sources. The lighting designer finds himself faced with the decision as to how much colour distortion is acceptable in the interests of efficiency.

Many of the decisions about colour of light source which the designer has to make have to be based on scientific knowledge which is as yet incomplete. Science cannot say precisely why one

colour of light source is *preferred* to another. It is even more difficult to explain why it is that one light source will be preferred in
some circumstances and at some levels of illumination, while
another light source, which is gloomy and unpleasant at low
levels of illumination, is bright, clean, and clinical at high levels
of illumination. Some of these things may be due to association.
Reddish light, low levels of illumination, and snugness and comfort may go together through association with firelight, a symbol
of warmth and security throughout the ages. Crisp bluish white
light, when there is plenty of it, seems to simulate best the character of natural daylight. Some of the problems of lighting and
vision can be specified precisely both from theory and experiment, while others can be specified precisely but on empirical
grounds alone, and yet others can only be handled numerically
on a sketchy and unsatisfactory basis. In every case, however, the
criterion of success is the satisfaction of the human observer.

VISUAL ACUITY AND CONTRAST

It is possible to express in precise numerical terms the amount of
illumination which is necessary on a visual task in order to permit detail of a given apparent size to be seen with a given certainty. In poor lighting the fine print of a newspaper cannot be
read, but the headlines may be legible. In average artificial lighting most people can read newsprint without difficulty, but they
need daylight or artificial lighting of comparable strength to daylight to handle matters of greater visual difficulty such as
threading a fine needle.

The numerical measure of the finest critical detail which can
be seen is usually expressed in terms of what is known as 'visual
acuity'. Visual acuity is defined as the ratio of the size of the
detail to the distance from the eye. In other words, acuity is a
measure of the angle subtended at the eye by the detail. It has
been shown experimentally by a great many research workers
that there is a clear numerical relationship between the magnitude of this angular subtense of detail, and the illumination on

the detail, the higher the level of illumination, the smaller the detail which can be seen with certainty.

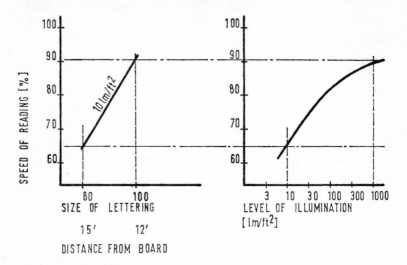

Fig. 6. Visual acuity can be improved by increases in the size and sometimes in the contrastiness of what we are looking at, as well as by raising the level of illumination. The diagrams show that at 10 lm/ft² a 25% increase in the effective size of, say, the lettering on a chalkboard will effect the same improvement as by moving from illumination at 10 lm/ft² into full daylight. (After Hopkinson, *Lighting* (H.M.S.O.), Part II, Section II, 1, Studies of Lighting and Vision in Schools.)

In the same way, it has been shown experimentally that the amount of light falling upon a visual task also governs the least difference in brightness between adjacent areas in the task which can be detected by the eye. The eye can see finer shades of brightness difference, or contrast, the greater the amount of illumination falling upon the task.

The sense of colour is also enhanced by the lighting level. Smaller differences of lightness or colourfulness of the same hue, and smaller differences of hue also, can be seen the higher the level of illumination.

Hence seeing is aided by higher levels of lighting in a threefold manner; finer detail can be seen, smaller difference of brightness or contrast can be recognised, and colour can be distinguished with greater certainty.

If therefore follows that it is possible to analyse any particular visual task in terms of critical detail, critical brightness contrast, and critical colour difference, and with the aid of the experimental relationships between these factors and illumination level, the amount of light can be predicted which is necessary to ensure adequate vision for any visual task. This has been done, and as a result tables of recommended lighting levels have been drawn up by the various responsible authorities to indicate the amount of light which is considered to be desirable for the adequate performance of these different visual tasks. Such tables of recommended illumination give the lighting designer a criterion for the quantity of light which he must provide in any given environment.

RECOMMENDED LEVELS OF ILLUMINATION

Various bodies speaking with authority on matters of lighting issue tables of recommended illumination level, or in some cases statutory requirements are made by national and local government bodies for minimum levels of illumination in certain specified buildings. It is usual to find considerable differences between the statutory requirements for minimum levels on the one hand and levels put forward for recommended practice on the other. The two serve a different purpose and it is important to recognise this. Minimum levels should never be used as a design criterion, but should always be exceeded because they are laid down on the basis that if the lighting does not reach this level, it will be definitely unsatisfactory for some reason or another.

A typical 'discrepancy' relates to the lighting of schools. The Department of Education and Science Building Regulations lay down that the minimum level of illumination in a teaching area should not be less than $10 \, \text{lm/ft}^2$, whereas other bodies, such as the Illuminating Engineering Society, in its Code of Recommended

Lighting Practice, recommends much higher levels. The Department of Education and Science, drawing on researches on the vision of school children conducted at the Building Research Station, established that if the level of illumination in a school classroom did not reach 10 lm/ft², normally sighted children would experience some visual handicap. As a compromise with economic circumstances, therefore, this level was a sound proposal for a minimum standard. On the other hand, the same work at the Building Research Station showed that higher levels of illumination result in a higher standard of vision, that is, better discrimination of detail and of contrast, but that the kind of visual tasks undertaken in schools did not normally require this higher standard of vision. Nevertheless if a higher level of illumination could be provided within the cost limits for the school, this could be nothing but advantage provided that unwanted effects, such as glare, were not introduced in the process. Indeed, the regulations make it clear that work of special visual difficulty, such as needlework, should have this additional lighting.

The use of tables of illumination giving different recommended levels for different visual tasks served a very useful purpose when artificial lighting was expensive because it influenced designers and management to provide adequate lighting. With the introduction of the fluorescent lamp, with its high luminous efficiency, and with the development of methods of electrical power distribution in buildings combined with the lighting 'trunking systems', it is simple and reasonably economical to provide levels of illumination in working areas which are well beyond the levels which were once recommended as good practice in the days of filament lighting. As a result, forward looking authorities such as the Illuminating Engineering Society in Great Britain are examining proposals for a new form of Code of Recommended Lighting Practice which takes into account the whole surroundings as well as the lighting level on the visual task itself. It is therefore likely that in a few years tables of recommended lighting level will cease to have the overriding influence in lighting design which they at present have.

THE WORK AND THE SURROUNDINGS

Before artificial light was cheap, work had necessarily to be done close to the artificial light source because there was insufficient illumination anywhere else in the surroundings. Later, general lighting of the whole environment became economically possible. General lighting of the whole building is at present the usual practice both by artificial lighting and also, of course, by natural lighting. Nevertheless many people feel an instinctive preference for some selective light on their own work. In buildings with good natural daylight, people prefer to work near or under a window. In artificial lighting, they like to supplement the general lighting by an individual light of their own which they can adjust to their personal needs. This desire for a personal light is by no means universal. It depends upon the kind of task which is being undertaken, but generally it is true for those visual tasks which demand concentrated attention on matters of fine visual detail.

This preference for individual light is in line with what we know about human vision. Experimental work has demonstrated that the best vision results when the work is slightly brighter than its surroundings, and that the greatest visual comfort results when there is a gradation of brightness from the visual task itself as the brightest area in the visual field, through intermediate zones to a moderate level of brightness in the general surroundings. Finally, it has also been shown that the attention is held most easily and distraction is avoided if the work is illuminated to a preferentially high brightness so that the eyes are drawn to it naturally and without strain. Experience and experiment, therefore, both lead to the decision that the best form of lighting ensures good general lighting in the building and in addition some preferential lighting directed for a particular purpose on to particular areas to aid in holding attention on the work and to improve visual acuity. This preferential lighting on the work does not necessarily entail the use of individual light fittings. It can be achieved by a suitable distribution of light from sources integrated in the fabric of the building. In cases other than working areas, special lighting may be in the form of emphasis lighting to direct the attention, such as to the display areas of an art gallery, to the chalkboard of a school classroom, to the goods on sale in a supermarket, or to the special religious features such as the altar in a church.

The building lighting and the special lighting may come from the same system, or they may come from different systems. During daylight hours it is usually much more convenient to provide the building lighting by natural means, while providing the emphasis light by artificial means. At night time there will be a combination lighting system, part providing the general building lighting while a special supplement, probably of directional characteristics, provides the emphasis lighting.

Experimental studies have shown that the best compromise between visual comfort and efficiency on the one hand, and excessively bright lighting on the other, is achieved if the illumination on the work is not greater than that which would render the work of an average brightness 'luminance' of the order of 1000 asb (100 ft-L). This figure for the maximum comfortable brightness in a working interior has been arrived at independently by different investigators and seems to be the figure at which about 20% of people with normal vision complain of too much light, the other 80%, of course, being satisfied. This seems to be a sound figure to use for guidance.

Comfort also depends upon the distribution of brightness. It is usually preferable for the work to be a little brighter than the surroundings. It has also been shown, however, that the eye is not very sensitive to the exact ratio of brightness of the work to the surroundings, but does become progressively more sensitive the higher the general level of brightness. Maximum comfort appears to arise if the general surroundings in the room as seen by the worker at his task are between one-third and one-tenth of the brightness of the task, that is between 300 asb (30 ft-L) and 100 asb (10 ft-L) and that there is an advantage if the brightness of the work is graded through an intermediate surround (for example, the desk top) into the general surroundings such that the ratio of brightness (luminance) of the work: immediate surround: general surround is of the order of 10 : 3 : 1 (Fig. 7).

These recommendations, or some modification of them, will probably form the basis for future lighting codes of recommended practice. It will be appreciated that there is relatively little room for wide variation in lighting levels at least so far as work places are concerned. In fact, it is very likely that in the future building interiors for work places will be designed so that the general building lighting will be of the order of 500 lux (50 lm/ft²), with preferential lighting for areas which require special emphasis, such as work areas, the amount of this preferential lighting depending upon the nature of the work or the degree of emphasis required. Lighting design in the future will therefore be much more a matter of luminance design. Selective tables of illumination level for different visual tasks will be limited to those tasks which require special treatment because of their particular visual difficulty.

DISCOMFORT AND GLARE

The eye cannot tolerate excess of light. While inadequate light leads to 'eye strain', discomfort and distress, too much light leads to glare and dazzle, and consequently discomfort of a different kind. The designer has therefore to steer a middle course between lighting which is inadequate for its purpose, and lighting which taxes the adaptation mechanism beyond its comfortable limits. Under daylighting conditions, glare results from a very bright sky seen through a large window. Under artificial lighting conditions, glare arises through a direct view of excessively bright light sources inadequately screened.

Glare is also a function of contrast. If a bright light is seen in dark surroundings, it will cause more glare than if seen in light surroundings. The juxtaposition of a bright area of sky and a dark wall can be intensely dazzling, even though out of doors the same sky can be looked at with impunity. There are consequently two aspects of glare which have to be watched in interior lighting design, glare which arises because of harsh contrasts between juxtaposed areas, and glare which arises because areas are of such an excessive brightness that the visual mechanism is saturated.

The main problem of glare in buildings is the avoidance of discomfort. It is rarely that glare in a building causes direct disability, although such situations do occur. For example, a window at the far end of an otherwise unlit corridor is a typical example of a design situation which gives rise to disability to vision due to glare. A high clerestory window giving on to the unobstructed sky can also cause disability in that areas around the clerestory cannot be clearly discerned, but normally it is discomfort which is the biggest disadvantage of clerestory lighting rather than disability.

Discomfort due to glare is not only a subject of complaint, but it is reasonable to suppose that it affects the general efficiency of the worker as a result of a build-up of annoyance, frustration and irritation in people who are subject over a long period to what amounts to a minor emotional affront. It has been shown, however, that the effect on human 'efficiency' is very difficult to measure, in much the same way as the effect of noise in a building is more important because of the distress which it causes than with the actual reduction of efficiency of working.

The relation between lighting and glare has been studied experimentally by methods which involve the precise evaluation of degrees of glare discomfort by skilled observers, resulting in the derivation of a Glare Index which is a measure of glare sensation. The Illuminating Engineering Society of Great Britain, drawing on research work at the Building Research Station, has drawn up tables of recommended limiting Glare Index for different environments. The experimental findings which showed that the probable degree of glare discomfort can be predicted from physical measurements of the important factors in the lighting of an interior, particularly the luminance and the size of the glaring light sources (artificial lighting fittings or windows), the average luminance of the surroundings which govern the adaptation

Fig. 7. Contrast grading. It is easier to concentrate on one's work if it is rather brighter than its background (see p. 55–7). For reading and writing the desk or table top should have a reflection factor of 20%–40% (Munsell value 5–7), and a matt surface.

level, the number of glaring light sources and their position in the field of view relative to the usual direction of viewing in the room (for example, in a school classroom of the more formal type the direction of the chalkboard might be taken as this usual direction). The experimental formula which links these physical factors to the glare sensation is complicated, but the evaluation was undertaken by computer as a result of which tables of the Glare Index have been drawn up and published by the Illuminating Engineering Society for artificial lighting situations, and by the Building Research Station for daylighting situations, so that it is possible to derive the Glare Index for an interior quite quickly from these tables.

The Illuminating Engineering Society, in its Code of Recommended Lighting Practice, tabulates these limiting values of Glare Index in parallel with the recommended levels of working illumination so that the two combined should give satisfactory lighting. The levels have been set as a result of extensive field studies, with the intention that if the Glare Index limit is not exceeded in a given environment, few, if any, complaints of glare should result.

BRIGHTNESS, COLOUR AND ADAPTATION

The part which colour plays in the design of the building interior is associated much more with the emotions than with the efficiency and comfort of the individual. This is one reason why colour problems are usually treated only very briefly in books dealing with lighting design. It is because the effect of colour upon the emotions, and particularly upon the degree of satisfaction and pleasure which colour can give in a building, are so little understood in quantitative terms that as a consequence design recommendations are often biased towards those limited aspects of colour which can be quantified. This is unfortunate but inevitable, and the situation can only be remedied when knowledge advances.

Our understanding of colour vision, and particularly colour vision in building interiors, therefore derives from two independent sources, the scientific sources which have established the fundamentals of colour vision, that is, how we distinguish colours in terms of their hue, their lightness, and their colourfulness, and on the other hand from non-scientific sources or partly scientific sources which give us some idea of the effect of colours on the emotions, on colour preferences, and similar effects obscure and difficult to define.

The physiological mechanism of colour vision is complicated and it would be inappropriate to discuss it even briefly here. The eye, however, can be shown experimentally to appreciate colour in terms of three basic attributes, the hue (that is, red, green, blue, etc.), the lightness or darkness of the colour, and the colourfulness. These three attributes were, for example, made the basis of Munsell's colour nomenclature, using the term Value for the lightness or reflectance, and Chroma for the colourfulness or saturation. In Munsell's nomenclature a colour with a high Value is very light, while a colour with a high Chroma is very colourful.

The appearance of a surface colour in a room will depend on its own properties and also on the properties of the light which falls upon it. A surface colour appears red because it reflects selectively wavelengths of light in that part of the spectrum which gives rise to the hue sensation of red. This is true because of the inherent properties of the surface to reflect red selectively this way, but the sensation received by the eye is dependent also on the fact that the surface receives white light, from the window or from the artificial lighting fitting, which itself contains a proportion of red radiation which can be reflected by the surface to the eye. If the incident illumination contained no red radiation, the surface could not reflect what was not there, and would consequently appear black. Consequently the appearance of any surface colour in the room depends not only upon its own reflecting properties, but also upon the nature of the light falling upon it. In normal circumstances such lighting will be inherently white, but there are differences in quality of so-called 'white' light which govern the appearance of colours.

As was mentioned earlier, the eye has a property of 'colour adaptation' which is of considerable importance in considering the effects of lighting by different kinds of artificial light source, particularly in considering comparisons between lighting by filament lamps and by fluorescent lamps, or by considering comparisons between either of these artificial light sources and natural daylight. It is due to this property of colour adaptation that a white surface appears white in the very different light of the filament lamp and of natural daylight, the one being relatively much more yellowish and the other much more bluish when compared the one with the other, but which have the property of allowing the eye to adapt completely to their very different types of light provided the one and not the other is present. In practice, we are hardly aware of the difference of the colour appearance of objects seen under filament lighting and under daylighting unless we are making careful colour judgments such as, for example, when choosing textile materials or when assessing the appearance of certain foodstuffs. Although the light of the ordinary incandescent filament lamp is very yellowish as compared with average daylight, very few people have any doubt that white paper looks white, whether seen by bluish daylight or by yellowish artificial light, unless the difference is pointed out to them in the form of a direct comparison. They judge white as white, and not as bluish white or as yellowish white, when they are fully adapted to the prevailing colour of the illumination. On the other hand, if there is the opportunity with the aid of some contrivance to compare the same white paper illuminated both by daylight and by artificial light, the very marked difference in the appearance is immediately obvious.

Different degrees of artificiality in the environment can, however, make even the layman aware of the fact that colour adaptation affects his life, and that this colour adaptation is sometimes insufficiently complete. The increasing use of greenish heat-absorbing glass in windows does occasionally make people aware of these colour adaptation effects. If one works in a room completely glazed by greenish heat-absorbing glass, the slightly greenish appearance of the interior of the room is noticed only when first coming in from the natural daylight outside. After a few minutes' adaptation, all sense of greenishness is lost, and the interior looks perfectly normal until one goes out again into the daylight outside, when there is an immediate sensation of wearing rose-coloured spectacles. This is due to the fact that the eye has been so conditioned by the greenish light that there is an after-effect which carries over for a few minutes but which is quickly lost and all seems natural again. The effect is even more disconcerting if it is possible to see through an open window.

Such colour adaptation is limited in its scope. If, for example, one works for a long period in a room lit entirely by red light (this is necessary practice in certain working environments, such as a photographic dark room), complete colour adaptation does not result and the sense of reddishness is not lost. Complete colour adaptation only takes place between different degrees of near-white light.

Colour adaptation, though not as important a factor in lighting design as is brightness adaptation, cannot be neglected entirely, particularly in view of the fact that new types of light source, such as fluorescent lighting in interiors, make demands on colour adaptation. In fact, the successful use of any particular form of fluorescent lamp will depend upon whether or not the eye can colour adapt to its particular form of radiation in such a way that complete satisfaction regarding the appreciation of surface colours in the interior is achieved.

While the eye is able to make considerable effort to adapt to the colour of a prevailing light source, it has certain definite limits. Colour adaptation cannot supply what is not there. The chief fault of certain forms of fluorescent lighting is that they are seriously deficient in red radiation. This deficiency is not noticed when looking at white paper because the blend of radiation coming from the fluorescent lamp is such that a sensation of whiteness, or near-whiteness, results. On the other hand, if the object (e.g. the human skin) depends for its colour characteristics upon its ability to reflect red radiation, the colour will be seriously altered if there is no red radiation present in the incident light. This change in colour may be seen as a change in hue, in lightness, or in colourfulness, or in all three. No amount of adaptation will result in any improvement. Whether the change in colour is of serious consequence will depend upon circumstances. The lack of red radiation in certain forms of fluorescent lighting results in changes in the colour appearance of the complexion; for example, the lips appear dark purple instead of red. In an office this can give rise to dissatisfaction and irritation, but no more. In a hospital recovery room the result may be that the onset of cyanosis in a patient recovering from an operation goes undetected, with the most serious consequences.

As a result of an understanding of the effects of colour adaptation, new forms of fluorescent lighting have been developed which do not show the same defects and which in particular contain a sufficient quantity of red radiation to permit the colour rendering of most objects seen in normal surroundings to be sufficiently close to that of daylight to cause little or no difficulty.

Brightness adaptation also affects the appreciation of colour. The appearance of a surface colour in a room will depend upon the amount of light which the surface receives relative to the general adaptation level in the room. A highlight, for example, a shaft of sunlight falling upon a patch of wall, will receive very much more illumination than the average in the room, and consequently the colour characteristics of this highlight will be modified. In general the effect of such selective illumination is to enhance colour, in the sense that it may appear lighter and more colourful. In the same way, a patch of colour in deep shadow will lose much of its colour attribute because it is seen at a very much lower brightness level than that of the average adaptation. The effect of shadow on colour appears to make relatively little difference to the judgment of hue, but results in a considerable reduction in the apparent lightness (Value) and colourfulness

(Chroma). A colour in shadow looks less colourful and darker, and so if a particular area, known to be in shadow, is required to have certain colour attributes, the effect of the shadowing can to some extent be compensated by the use of lighter and more saturated colours.

THE EFFECTS OF CONSTANCY

Constancy is the name given by psychologists to the human ability to see things as we know they should be, rather than as they are. Constancy manifests itself in various ways. The classical demonstration of 'size constancy' is that of the three men seen against a grid of converging lines, which is interpreted by the eye as a perspective grid leading to a point at infinity. As a result, the three men, who are drawn exactly the same size on the sheet of paper, appear because of size constancy effects to be of greatly different size. These constancy effects in the appreciation of space in an interior have been demonstrated in exaggerated form with the aid of distorted perspective model rooms, such as the Ames Room. Lighting can contribute to the exaggeration or the diminution of the size constancy effects. The exuberance of solid plaster work combined with painted illusions in Baroque interiors is one example of the putting into effect for interest and amusement of the constancy characteristics of human vision, that is, the desire to see things as we think they should be.

Brightness constancy is always present in a lighted interior, although its effects are rarely appreciated without being pointed out. The light falling on a ceiling in a room lit by side windows may be, at the back of the room, only one twentieth of the amount of light falling on the ceiling near the window, yet the eye sees the ceiling as white all over, rather as than white near the window and as dark grey remote from the window. It is easy to demonstrate that the physical brightness of a dark grey surface placed on the ceiling near the window may be as high as that of the white ceiling at the back of the room, yet this is not how it appears to the eye. We know from experience that the ceiling is

5

of a uniform white reflectance and so we discount the fact that all areas of the ceiling do not receive the same illumination. This is only true provided we can see the whole ceiling and the window. Seeing them both, we are then able to recognise that light reaches the ceiling from the window and from nowhere else. Thence we deduce that more light reaches the ceiling near the window than at the back of the room. In judging the appearance of the ceiling we make allowance for this, we say to ourselves: 'This is a uniform white surface; it is receiving a lot of light near the window and it is receiving little light away from the window. Therefore it appears not as graded from white to dark grey, but as uniformly white all over.' If we were to view a small patch of of the ceiling through a hole in a black box placed over the head, it would be impossible to tell that the ceiling at the back of the room was a white area receiving only a little light, or a grey area receiving much light. Brightness constancy only operates if we are able to take in the whole visual scene, including the source of light as well as the surface receiving light. In these circumstances we can distinguish between the illumination component and the surface reflectance component of the brightness of a surface.

In the same way, our eyes recognise colour for what it is, even when colour is seen under the influence of coloured light. If a room is provided with a highly colourful red floor covering, reddish light will be reflected from the floor to the ceiling, and give the ceiling a distinctly pinkish hue. If, however, we look at the ceiling, and at the same time are able to appreciate that the floor is a brilliant red, the pinkishness of the ceiling will be sublimed in our consciousness, and it will look white. On the other hand, if our view of the floor is completely cut off, and we have no clue whatsoever as to what is causing the pink light on the ceiling, the ceiling will no longer look white, but pinkish. A less extreme example is the greenish light reflected from a lawn outside the window bathed in sunlight. Upon close examination, once we are aware of the fact, it will be seen that the ceiling near the window receiving direct reflected light from the green lawn outside is, indeed, quite greenish.

Both brightness constancy and colour constancy have an effect on the judgment of the appearance of things seen, and any attempt to design for the appearance of an interior must take both effects into account, either implicitly, as is usual, or, more rarely, explicitly. These are effects which the skilled designer of interior decoration knows how to allow for even though he may make this allowance qualitatively rather than quantitatively.

4

Natural Lighting

DAYLIGHT AS AN AMENITY

The development of cheap artificial lighting in the form of fluorescent tubes has raised a doubt for the first time as to whether daylight is really necessary. Why not exclude the fickle natural environment, the argument goes, and provide a stable, controlled, man-made environment in which the lighting, the heating, the air conditioning, and the rest are all under careful and constant control of a building supervisor. These arguments are becoming more and more vocal and they should be examined with some care.

It has been stated that no evidence has yet been produced to show that any harm to sight or to general well-being results from continuous work in artificial lighting. This is the form in which the statement is made by those who are concerned, for good and sufficient reasons, to promote the wider use of artificial lighting in buildings. It is not unreasonable, however, to ask that the argument should be put the other way, and to demand evidence to show that any gain in vision or health results from working in a controlled man-made environment, and in particular to show whether there are any advantages in working in continuous artificial lighting as compared to natural lighting.

When examined in this way, the evidence, such as it is, favours natural lighting. No positive gains have ever been produced on health grounds for artificial lighting. During the past twenty years the standards of artificial lighting in school buildings, for example, are very much higher than they were forty years ago, yet the onset of juvenile visual defects is no less now than it was.

This statement makes full allowance for earlier and more accurate diagnosis at the present time.

Deprivation of daylight, on the other hand, is believed sometimes to cause health disorders, either 'somatic', that is, the upset of the regular bodily functions, or 'psychosomatic', that is, nervous upsets of various kinds leading to illness and ill-health. The evidence for this comes, as we might expect, from northern countries—Sweden, Finland and Russia, where there is a serious lack of daylight in the winter months. The evidence is not conclusive, because it could well be argued that these illnesses arise from factors other than the lack of daylight, such as the separation of workers in the northern mines and forests of Sweden from their families in the south. Nevertheless these same illnesses do not happen in the same way in other climatic environments where daylight deprivation is not a feature.

People can, however, work in windowless environments apparently with little effect on their health. It seems that people will accept lack of daylight when they know that for very good reasons they cannot have it. People will work in large department stores, in underground railways, and in factories where there is obviously a valid reason for a windowless, enclosed, and insulated building, such as the need to control the temperature and humidity for a delicate industrial process. Any resentment people feel against a lack of daylight seems to be against deprivation of an amenity to which they feel they are entitled and which they could reasonably expect to have.

Even so, in the present state of evidence no designer free of commercial and economic constraints would feel confident in designing a building from which all daylight is excluded merely on the grounds of economy or convenience in design. When other amenities are equal, staff change—when they can—to a job with daylight. The case in favour of the man-made environment from which daylight is excluded needs considerably more evidence. This is particularly so in temperate climates where the natural environment is often comfortable and the outside world contains much that is visually pleasant and interesting. Things may well be different, however, in an unfavourable natural environment.

The advantages of daylight over artificial lighting must be sought in the difference between them. Above all, daylight is variable in quantity and in colour continuously throughout the day, except on fully overcast days, when the general dulling and depressing effect which most people experience on such days offers a significant argument against the invariance of artificial lighting. Daylight has a different spectral composition from any form of artificial light and while the colour appearance and to some extent the rendition of colours given by daylight can be simulated artificially, the simulation is not perfect. Vegetable and animal physiology has revealed that the effects of light on an organism often depend upon the presence of spectral radiation distributions identical in every respect, and not merely upon the presence of certain dominant spectral bands.

Probably the real advantages of daylight, however, are to be sought not so much in the spectral composition of the light, though that is important, nor in the variability, though that may well be important too, but more in the fact that daylight is provided by windows and windows have the dual function not only of admitting light, but also enabling us to see outside and to maintain visual contact with the outside world during the working day. Teaching staff often say that the pleasure of working in a post-war school is not only that the light is good and adequate all over the classroom but also that the children are continually aware of the changing pattern of the world outside. They can see that the clouds move, that the moon can be seen in the blue sky during the day, that the rain comes from heavy billowing groups of clouds, and that, in fact, the external world is continually changing in an interesting and fascinating way. These are things that some older school administrators consider to be undesirable distractions, but they are now found to be pleasant intermissions from which the children return refreshed to their work in hand. If this is true of children, it may well be true of adult workers also. Certainly there is considerable evidence that one of the most important functions of a window is to provide a visual rest centre, an amenity which provides a link with the outside world and the different facets of the passing day. There seems to be a real human need for variety and change and windows offer the physical and psychological relaxation of being able to gaze out at distant objects without the muscular effort of visual accommodation.

The case for providing daylighting in buildings is not well documented because it has never had to be put until now. Up till near the present time, daylighting was the only economical means of lighting a building. Now a decision has to be made on a fundamental matter and there is insufficient evidence to guide the choice because there has never been the need to collect such evidence. The next phase in daylight research will have to be concerned with such matters and the evidence will have to be collected quickly because the economic pressures are growing in favour of the more extensive use of permanent artificial lighting. There is no doubt that people judge the adequacy of the daylighting in a building first and foremost by whether they can see to do their work well and without visual strain. If they have more than sufficient daylighting to see well, they then judge by the brightness of their surroundings, particularly the view of the sky, and then by whether the walls in their visual field look bright and well illuminated. Then if they are particularly critical, they judge the daylighting by the shadows and by the presence of dark areas and by the limitations which these dark areas place on their ability to locate their work around the room.

But above all their judgments are governed by the amount of bright sky which they can see. This bright sky can have a dual effect. If it is directly in their field of view, and they cannot turn away from it, it can raise the adaptation level to the extent that areas in the depth of the room, though receiving plenty of light, nevertheless look apparently dark and are so judged. If the bright sky is not directly and continuously in the field of view, the effect can be pleasing.

The desire for sunlight is another matter. Man has a love-hate relationship with the sun which is never more apparent than in the interiors of buildings. The balance of love and hate is quite different in the temperate maritime climate of Great Britain than, for example, in the dry southern areas of the United States. Very few people in Great Britain would wish to have sunlight totally and permanently excluded from the places in which they live and work. Nevertheless they do not want sunlight all the time and if they are to have it, they would like to have it under their own control. In the past there have been two ways of approaching this problem of wanted and unwanted sunlight. In the course of an extensive social survey conducted some twenty years ago, housewives indicated that in general they like to have sunlight in the places where they are living and working, particularly in the morning, but that they are worried about having too much of it, particularly in living rooms where excessive sunlight can cause fading of furnishings and overheating of the living space. The survey also revealed that people tend to say that they want sunlight in the places where investigation reveals that they already have it.

Another investigation used an entirely different approach. A group of people were asked to write brief essays giving the reasons why they like to have sunlight in their homes: one of the interesting comments was that sunlight penetrating into a room creates a tangible link with the world outside. Sunlight was liked for its ability to enhance colour and so to give a sensation of excitement and uplift. Sunlight also has the more prosaic property of increasing the amount of working light in a room both directly and by inter-reflection. For whatever reason, and there are many, people do like sunlight in their living and working places provided that they do not have to endure too much of it for too long and particularly that it is not allowed to make their working conditions uncomfortably hot.

DAYLIGHTING FUNDAMENTAL TO
BUILDING DESIGN

The decisions that the architect and his client make about the part which daylight is to play will have the most far-reaching effects upon the building which they are creating. This may seem self-evident, but many new buildings appear to have been designed with priority given to other less relevant considerations. The size and position of the windows seem to be determined more by their outward appearance than by their effect upon the environment within the building. Many architects seem still to be engaged in modern skirmishes in 'the academic wars of fenestration—marching and counter-marching of the formal Malbroukian type' of which McGrath writes in *Glass in Architecture and Decoration*—although he had a different period of architecture in mind. So there have been in turn the glossy packages of glass curtain walling; the heavily rusticated blocks with each window boxed out separately; the elongated vertical slits spaced at random down the sides of libraries, monasteries, and town halls. In each the justification for the form the daylighting takes is often either absent or supported by shaky rationalizations.

The outward appearance of a building is, of course, of prime importance and indeed may be the only aspect of it that can be appreciated by those who pass by. Nevertheless the external appearance must never be achieved at the expense of the well-being of those who live or work in the building. The achievement of a satisfactory relationship between form and function is one of the main tenets of modern architecture but it is one that is more honoured in the breach than in the observance. In the most successful new buildings a stimulating new aesthetic emerges as an expression of new techniques used in a way that satisfies human needs, which are now understood much better than ever before.

Since one of these fundamental human needs is for access to daylight and to the sun, the architect, when considering the general form and character of a building, will have to bear in mind this requirement for daylight at the same time as he tackles the design of the space provided for the activities within the building or of the physical environment needed by its occupants.

Thus in the course of this process of design for, say, a school, daylighting to a given minimum level and with certain other characteristics will be required. This will indicate a range of solutions in plan and section, within which the architect may work. It will be found that working positions cannot be more than a certain distance from windows, unless rooflights can be introduced. This in turn may suggest that large rooms should be in single-storey construction or be on the top floor of multi-storey blocks. The architect must then decide whether the smaller rooms should have windows in one wall only, weighing in his mind the limitations of unilateral lighting against the economic advantages of the compact double-banked multi-storey block which this permits, or whether the building should have a longer coastline to allow for windows on two or more sides with the improved quality in lighting and the greater freedom in use which this gives. The ways in which the skill of the architect designing a school is tested in finding solutions which meet both the economic and the educational requirements of his brief is discussed later. Here it is only necessary to emphasise that decisions about daylighting requirements have fundamental implications not only for the layout, orientation, and appearance of our buildings both by day and night, but also for the principles of town planning on which much of the character of our cities depends. Considerations of daylighting are a vital part of town design not only because of the effect each building has upon the light available to its neighbours, but also because ideally the views out of its windows should be designed as carefully as the amount of daylight within the building.

To insist upon daylight for all working illumination—particularly for large buildings on urban sites—is now unnecessary, impracticable, and even uneconomic, since modern techniques in lighting permit the use of specially-designed artificial lighting as a permanent supplement to daylight during the working day.

This technique of permanent supplementary artificial lighting in interiors (see Chapter 6) has now become part of regular lighting technology. It need only be pointed out here that in such buildings their aspect, orientation and the views from them, together with the placing and the detailed design of the windows, will need to be considered with particular care. The main advantage that is claimed for this approach, other than its lighting qualities, is that the smaller windows will reduce both solar heat gain during the summer, and thermal losses from the building in winter. At the same time, however, a reduction in window area and higher levels of artificial lighting will, in deep buildings, intensify the problem of ventilation and cooling during the summer.

The role of daylighting in a building may conveniently be considered in two broad categories, first, the ways in which it helps people to see well, and second, the contribution which it makes to the amenity and general character of the building.

In order to provide daylighting which is appropriate to the function of the building, the architect must first have a thorough knowledge and a sympathetic understanding of how the building is going to be used. He must get his client to explain in detail what activities will be carried out in what places and at what times. The visual requirements should emerge as part of the general design brief to the architect. All too often the client will not be sufficiently skilled at explaining his real needs and it is often up to the architect to extract the necessary information by close questioning. This vital part of an architect's job, often ignored in schools of architecture, cannot be sufficiently stressed. It is usually worth visiting the client's present accommodation if he has any or otherwise visit existing buildings of a similar type with him. One often finds that the client's preconceptions about daylighting may spring from something which he has seen or something to which he has become accustomed and which may not be suitable to his present needs.

A clear brief about the visual tasks for which the architect is to provide will then indicate the levels of daylighting required. An analysis in terms of the critical detail and critical contrast of the tasks may be made but more usually the architect can refer to the various codes of good lighting practice such as those of the British Standards Institution or the Illuminating Engineering Society. These publications generalise the visual requirements of different building types and make recommendations on what the levels of working light should be.

Useful as these publications may be, they do not serve as a substitute for clear thinking about the problems from first principles. Visual tasks are often thought of as being those demanding immediate co-ordination of hand and eye such as the operation of a lathe or the dissection of a specimen in a biology laboratory. Equally important, however, may be the ability to see other people easily: the student watching the lecturer's face, the foreman supervising an operative, the nurse studying her patient, a dress buyer appraising a model at the spring collection. Each will require to be able to see detail, colour, and movement clearly and without distraction. In some buildings the visual tasks are not limited to within the building. They may involve enabling a passer-by to see inside as in a shop, or deliberately preventing him from looking into a ground floor flat, although the occupants may themselves still wish to be able to see out.

It is important for daylight design to know the frequency with which different tasks will be done, whether or not they are fixed in location, and the numbers of people who may be involved at any one time. For example, if an individual worker has an intricate job to do, it may be reasonable to suppose that he can take it to the window. If, on the other hand, twenty biology students have simultaneously to carry out a detailed dissection, then an adequately high level of working light must prevail over the whole of their laboratory. For this reason the most severe daylighting problems for the architect are in modern buildings such as schools, big offices, and factories where large numbers of people may all be doing realtively difficult visual tasks at the same time.

An analysis of the distribution of visual tasks in a room will have to indicate whether particular rooms need to be lit to a high level throughout or whether high levels are needed only in certain areas with a lower level in the rest of the space.

LEVELS OF ILLUMINATION FROM DAYLIGHT—THE DAYLIGHT FACTOR

For some building types such as schools and factories there are recommended levels of daylight illumination. These are given in terms of the Daylight Factor. The Daylight Factor has been found to be the most useful and appropriate measure of the daylight in a building, particularly in the overcast maritime climate of north-west Europe. It expresses the total interior illumination, that is, the light which reaches the interior directly from the sky, by reflection from outside obstructions, and by inter-reflection from the surfaces within the room, as a percentage of that available outdoors. This is a valuable concept for two reasons. First, there is the adaptation effect described in Chapter 3. The eye has powers of adaptation which to a great extent, though not completely, compensate for the variation in the daylight, and so a ratio of the interior to the exterior light is a measure which to some extent takes into account this adaptation effect. Second, the Daylight Factor is an arithmetic convenience because it is not only a measure of the lighting which accords well enough with the visual effect on the eye, but it is also a quantity which can be expressed in terms of the size and position of the windows relative to the reference point, the reflectance of the internal and exterior surfaces, the angular obstructions of these exterior surfaces, all of which are measurable physical quantities which can be fed into an appropriate computing formula to enable the Daylight Factor to be calculated with the design of the building.

The recommended levels of Daylight Factor are arrived at by a process of reasoning of which the two main considerations are (a) that on dull days a sufficient amount of working daylight will be provided everywhere it is needed, and (b) that this sufficient level will be provided over the greater part of the working period throughout the year. In Great Britain it is often taken that a 'dull day' in the above context is that which gives a total illumination from the whole unobstructed sky of 5000 lux or 500 lm/ft², and 'the greater part of the working year' is taken as 90% of the time from 9 a.m. to 5.30 p.m. throughout the year. The recommended levels of Daylight Factor published by the British Standards Institution, the Illuminating Engineering Society, and the statutory minimum levels of Daylight Factor laid down by the various building Government Ministries, have all been determined from this basis. (See *Daylighting*, Hopkinson, Petherbridge and Longmore; Heinemann 1966.)

The first stage of the design process is therefore to establish the necessary levels of Daylight Factor, by reference to the appropriate authorities, or more rarely from first principles.

It is usually then possible at this stage in the design process to establish the general form and method of achieving the daylighting, or at least to fix two or three likely alternatives and so to proceed to the detailed calculation of the Daylight Factor. Before this is done, however, some consideration must be given to other aspects of the user's requirements in addition to that of the quantity of daylight illumination. It is not enough for the daylighting merely to provide the right quantity of illumination. Good design must also allow the occupants to see without strain and at the same time help to establish an appropriate character for the interior. Visual comfort in daylighting depends largely on an absence of glare and upon the right distribution of brightness throughout the room.

REVEALING ARCHITECTURE

Daylighting in a building has the primary function of enabling people to see. However, the architect is also concerned with the way in which the solution to this lighting problem interacts with other aspects of the building. It has already been observed that the method of daylighting and the distribution of the windows has a far-reaching effect upon the layout and form of the building both in plan and section. Again, the visual character of the interior is primarily dependent upon the means by which daylight is brought into the building and the way in which the interplay of light and shade is used to reveal form, surface, and space.

The secluded view out of a small domestic window or the inviting sweep of a fully glazed shop front each have their place, and each helps to set the tone of the building both from inside and from outside. In a different way the proportions of direct and indirect daylight will have a marked effect upon its character. A high proportion of direct light will give strong modelling and a rather heavy dramatic character. On the other hand, a room with a high proportion of indirect light will have lower contrasts and a softer, more restful character. The decisions about what is the most suitable character—inviting, solemn, gay, or restful—will depend upon an imaginative interpretation of the uses to which the room is to be put.

The view outside through the window is also a factor in determining the character in a room. Windows must be spaced and placed so that not only can one see out easily from all normal positions in the room but that windows should, if possible, give a natural view of an interesting portion of the scene outside. Exaggerated vertical and horizontal slits restrict the view in an unnatural way. Tradition and familiarity have a great deal of influence in this, but it seems that many people prefer a window which is slightly wider than its height because normal eye movements tend to be more frequent to and fro along the horizontal than up and down the vertical. The best shape for a window as a frame for a view will, of course, depend to some extent on the nature of the view. Window bars which break up the view can be an annoying distraction, though they will be less troublesome if they do not introduce a harsh contrast in brightness. A horizontal bar at or slightly below eye level is particularly distracting.

DAYLIGHT DESIGN—THE SOURCE

The ultimate source of daylight is, of course, the sun, but it is sunlight diffused by the sky which supplies the major component of interior daylighting in the temperate, overcast climate of north-west Europe.

The relative roles played by direct sunlight and by skylight in interior lighting depend upon the locality, that is, the latitude on the earth's surface, and also upon the local climate with respect to frequency of cloud cover. These two factors determine the building traditions in relation to windows and the grouping of buildings. In the crowded cities of Greece and Spain, and in sunny climates generally, much of the interior lighting comes through small windows after reflection on the walls of courtyards and opposing façades. Much of this light is reflected sunlight; the sky is often a deep blue of low brightness and so provides relatively little of the total amount of light reaching the interior. In north-west Europe, on the other hand, the sun is seen sufficiently rarely to be unreliable as the main source of interior daylight, and so design is based on the light of the sky, resulting in large windows giving out on the uninterrupted sky wherever freedom from obstructions make this possible.

In Great Britain and neighbouring countries, the light of the sky varies in quantity over the hours of the day and the seasons of the year quite apart from the unpredictable changes caused by local weather conditions. On the other hand, in sunny, dry climates the light of the sky remains fairly constant throughout the day and the only changes in daylighting which occur are those brought about by the change in position of the sun in the sky.

Whether interior daylight is planned for appearance or for the reading of small print, it can only be planned at all by making certain assumptions about the brightness of the sky. These assumptions can be made on the basis of average sky conditions, or they can be based on limiting conditions which need not necessarily be either maximum or minimum conditions. The height of doors is not based on the average height of the human race, nor upon the height of the tallest man, but upon an agreed value which will cause the minimum of inconvenience consistent with structural and economic considerations. In the same way daylight design is based on sky conditions which are agreed as representing a reasonable compromise between all the various factors involved, paying particular attention to the prevailing climatic conditions and the habits of the people who are going to use the building.

International discussion between lighting authorities has resulted in agreement to adopt two standard conditions of skylight distribution for application to daylighting design. These two conditions are (a) the international standard overcast sky, and (b) the uniform brightness sky. The standard overcast sky serves daylight design in all maritime, cloudy, temperate regions, and is based upon extensive measurements of sky brightness (luminance) distribution under conditions where the sky is so fully overcast that the position of the sun cannot be seen. The sky of uniform brightness (luminance) on the other hand is an idealised condition, which never happens in reality, but its use leads to many simplifications in daylight calculation (less significant now with computing facilities readily available than originally when it was first adopted). The blue sky in dry sunny climates where clear atmospheres prevail has a brightness distribution over the greater part of the sky, except the area near the position of the sun, which can be taken as of uniform luminance, and so the uniform luminance sky is often adopted for daylight design in such dry, sunny regions of the world (see Chapter 12). The luminance distribution of the international standard overcast sky (the CIE sky, so called from the Commission Internationale de l'Eclairage) is not uniform although it is symmetrical about the zenith. Such a sky is darker at the horizon, the luminance distribution being given by the formula:

$$B_\theta = \frac{B_Z}{(1 + 2 \sin \theta)}$$

6

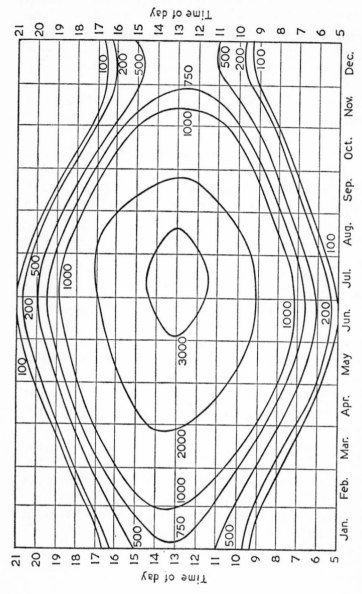

Fig. 8. Variation of given levels of sky illumination (lumens per sq. ft.) with time throughout year (based on measurements at Teddington, Middlesex). (After Fig. 3, *Daylighting*, B.S. Code of Practice CP3: Chapter 1: Part 1, 1964.)

where B_θ is the luminance at an angle of elevation θ above the horizon, and B_z is the luminance of the sky at the zenith.

Thus for the horizon, where $\theta = 0$, and $\sin \theta = 1$,

$$B_\theta = \frac{B_z}{3}$$

that is, the luminance at the horizon is one-third the luminance at the zenith.

The adoption of these two standardised sky luminance distributions permits daylight design to proceed on a firm mathematical basis, but it is still necessary to take into account the wide variations from the standard conditions which occur in practice.

Measurements of sky illumination made over a period of years have permitted data to be prepared of the variation of sky illumination to be expected throughout the year (Figs. 8 and 9). Such tables of available daylight have been published for a number of areas in the world.

		Illumination (lm/ft²)						
		10	15	20	30	45	70	100
Daylight Factor (per cent)	0.5	35	5	—	—	—	—	—
	1.0	67	52	35	5	—	—	—
	2.0	85	75	67	52	24	—	—
	3.0	90	85	79	67	52	22	—
	5.0	93	91	88	81	71	54	35
	10.0	95	94	93	91	86	78	67

(Calculations based upon 9.0-5.30 p.m. five-day working week giving a yearly total of 2,186 hours.)

Fig. 9. Percentages of total working time of the number of hours during the year for which the illumination exceeds recommended levels at given daylight factors. (Reproduced with permission from I.E.S. Technical Report No. 4.)

The probable daylight to be expected in other areas can often be derived by extrapolation, taking, for example, the published relationships between probable levels of daylighting and the latitude. Again, where meteorological data of total sky radiation exist, these can be converted to sky illumination by means of the 'sky luminous efficiency factor'. This is a figure which gives the proportion of the total radiation which lies within the visible region of the spectrum (the value is of the order of 120 lumens per watt of total radiation). Systematic design of interior day-lighting is possible only for climates which are consistently sunny, or consistently overcast. It is then possible to compute the interior daylighting using either the uniform luminance sky, or the international standard overcast sky, as the basis for the computations. Areas like Britain experience wide variations in sky conditions, from the fully overcast sky, to the blue sky with white cumulus clouds, extremes of conditions which bracket the widest range to be found anywhere else in the world. Under such a wide range of sky conditions it is necessary to determine what is the critical design condition. In Britain it has been demon-strated that the overcast sky is the critical design condition, and so interior design in Great Britain is based upon the international standard overcast sky. It is true that in Britain the southern sky is more frequently brighter than the northern sky, but window design does not take this into account. If the south facing win-dows were made smaller, such rooms would consequently suffer on fully overcast, dull days, when the southern sky is no brighter than any other part of the sky, and when light is needed most.

In addition to the change in brightness which occurs with the changes in sun and cloud conditions, the colour of the light from the sky also changes. The direct light of the sun is warmer in ap-pearance than the overcast sky, which in turn is less blue than the clear sky. More precisely, these changes can be expressed in terms of the 'correlated colour temperature' which is a physical measure (related to the spectral distribution of radiation from a full radiator at the quoted temperature). The correlated colour temperature of the direct light of the sun, of the average overcast sky, and of the average blue sky respectively is often cited as 5500 °K, 6400 °K, and 11,500 °K. (Fig. 10) These differences in colour of daylight are important when it is necessary to consider the colour of an artificial light source to be used to supplement daylight in working interiors.

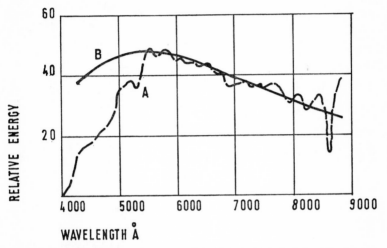

Fig. 10. Spectral energy distribution of daylight in Britain (after Henderson and Hodgkiss).

    A. Mean of 50 curves, 6400 °Kelvin

    B. Full radiator at 6500 °K

The appearance of colours under daylight is generally accepted as their normal appearance. This needs to be borne in mind when choosing colours and materials for an interior that will be seen under both daylight and artificial light, and in particular where both are to be blended (see Chapter 6).

DESIGNING FOR DAYLIGHT QUANTITY

The principles of daylight design derive from the physics of radiation transfer. The daylight reaching a point in an interior can be considered as a sum of (a) the direct light from the sky, (b) light reflected from external surfaces, and (c) light reflected and inter-reflected from internal surfaces.

    The direct sky component (SC) is usually the major part of the total daylight, particularly in an unobstructed situation.

The externally reflected component (ERC) is of significance only in heavily built up-areas.

The internally reflected component (IRC) includes all the light which reaches the reference point in a room after multiple internal reflections, that is, light from the sky reflected from the floor and thence from walls and ceiling, light from the ground outside the window reflected through the window to the ceiling and thence around the room, and so on.

At points near the window, the sky component (or the externally reflected component in a heavily built up area) will constitute the major portion of the total daylight, whereas at points remote from the window, the internally reflected component can account for a large proportion of the total daylight provided, the internal room surfaces are of high reflection factor (Fig. 11). At points in the room which have no direct view of the window, all the light reaching these points comes from internal reflections.

The division of the total daylight into three distinct components permits a clearer understanding of the significance of the parts played in daylight design by different building elements.

The windows govern the whole daylight, both sky and reflected components. The greater the area of glazing, the greater will be the amount of daylight admitted, but the amount reaching the reference point will depend not only on the size of the windows but on their position relative to the point. The illumination coming from a point source to any plane surface varies inversely as the square of the distance of the source from the point (the inverse square law) and also upon the tilt of the plane surface relative to the perpendicular direction of the source (the cosine law). If this angle of tilt is great, the light coming from the point will be spread over an area greater than if the surface was set at right angles to the direction of the source. The illumination in a plane is therefore proportional to $\cos \theta / D^2$, where D is the distance from the reference point to the source, and θ is the angle between the normal to the plane at the reference point and the direction of the source.

The combination of the inverse square law and the cosine law

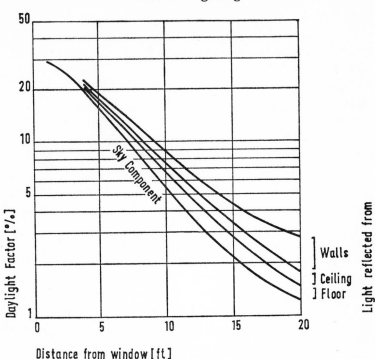

Fig. 11. A high proportion of the daylight factor at the back of single side lit rooms with light surfaces will be contributed by reflected light. In the example shown in the diagram, the average reflection factors are: ceiling, 70%; walls, 35%; floor, 20%.

of illumination has an obvious influence on window design. If we are considering light for a visual task on a horizontal surface, which is usually the case for living and working interiors, it follows that the higher the window above the reference plane, the greater will be the sky component, other things being equal, for the reference surface is then more nearly normal to the direction of the window. Thus the same area of glass placed in the roof above the reference point will contribute more light than if placed at the same distance away but in a vertical wall. A tall, vertical window will generally give deeper penetration of daylight on to a horizontal surface like a desk top than the same area of glass set into a low, horizontal-view window. A proper understanding of the simple laws of radiation transfer permit a precise calculation to be made of the relative efficiency of different window designs in providing illumination where it is wanted.

Losses of light through windows can cause a significant diminution in the daylight available indoors. Obvious reductions (obvious though often neglected) arise through the obstructions caused by window frames and glazing bars. Even in a light metal-framed window, the actual glass area may be only 75% of the area of the window aperture, while a heavy wooden frame can account for more than half of the window opening. Allowance must also be made for the loss of light due to the glass itself, even when clean. Dirt on the glass accounts for further losses, though the diffusing and obscuring effects make one aware of dirt on a window even when the actual light transmission has only been slightly affected. Internal structures, ventilation pipes and so on, placed near a window, all reduce the available internal daylight. If all these accountable losses are not included in a daylight calculation, it is easy for the probable daylight to be overestimated by a factor of 2 to 1 or more, with consequent disappointment by the client in the event of check measurements being made photometrically in the completed building.

OBSTRUCTIONS AND EXTERNAL SURFACES

Obstructions and external surfaces reduce the total amount of direct light coming into the room while offering in compensation only the small amount of light reflected from their surfaces. Where these external surfaces are under the control of the designer, which is rare indeed, they can be designed, cleaned, or painted to a high reflection factor to add to the externally reflected light and also, incidentally, to reduce the harsh contrast between sky and dark building.

DECORATIONS OF THE INTERNAL SURFACES

Decorations of the internal surfaces in a room have a major

influence on the daylight. It can be shown that the internally reflected component is determined by the formula:

$$IRC = \frac{\text{first reflected light flux}}{A(1-R)}$$

where A is the area of the room surfaces (walls, floor, and ceiling) and R their average reflectance. Clearly the greater the value of R, the greater will be the internally reflected component. In addition, the first reflected light flux on an internal surface is the product of its reflectance and of the direct light coming first to that surface through the window. The sum of all such first reflected light fluxes is taken for the formula above. Hence both the reflectances of individual surfaces and the average reflectance of the whole room are of importance in determining the internally reflected light. If quantity of light is the prime consideration, it follows that surfaces which receive the most skylight should consequently have the highest reflectance in order to make the greatest contribution to the total daylight in the room.

The floor receives the most direct skylight, and so it is most important that the floor should have highly reflecting properties.

The walls near the window also contribute significantly and should also have a high reflectance.

The ceiling and the back walls, on the other hand, receive little direct skylight. It could be argued that they need not have a high reflectance. This is true so far as the influence on the first reflected light is concerned. The advantage of giving back walls and ceiling a high reflectance is that they then *appear* bright. It is for subjective reasons primarily that they should be light. In addition, a light-coloured ceiling and a light-coloured back wall does enhance the internally reflected light by raising the average reflectance of the whole room so that it has quantitative significance as well as contributing to the sensation of lightness and freshness in the room.

The *window* and *reflected light* are related somewhat equivocally.

Glass does not reflect much light. The diffuse *reflectance* of a clean pane of glass is of the order of 15%, and so a large window reduces the average reflectance of the room and hence *reduces* the internally reflected light. A window lets direct skylight in, but it also lets reflected light out. A room with very large windows will therefore depend upon direct skylight for the interior illumination, whereas a room with windows of moderate size and with internal surface of high reflectance will have a blend of direct light from the sky and diffuse light reflected from the internal surfaces.

WINDOWS BELOW WORKING LEVEL

Windows below working level admit light to add to the reflected component in a room, although, of course, no direct light from the sky can reach the working level through such low windows. Such low level windows combined with a reflecting floor of light colour can contribute substantially both to the quantity of total daylight by reflection, and, by choice of colour of the reflecting floor surface, it can produce a suffusion of colour to add warmth to an otherwise austere room, or coolness to a room in which there may be too much warm colour, for example, from wood furniture.

DESIGNING FOR QUALITY— BRIGHTNESS AND GLARE

Comfortable vision in an interior results from a satisfactory balance of brightness across the room with no harsh contrasts and with no aggressive differences in brightness between one area and another to distract the attention. Light is both an attraction and a distraction and it has to be properly organised to provide exactly what is wanted.

The appreciation of brightness in a room is governed by adaptation and contrast. Areas appreciably below the adaptation reference level look dark, even if the actual physical quantity of light is not low. If the brightest visible surfaces in a room are confined to the areas near the windows, people sitting at a distance from the windows will find that their immediate surroundings appear gloomy, even though the actual quantity of daylight on their work may well be above the recommended statutory minimum. Such lighting may meet legal requirements, but it will not be good lighting. In order to correct such an imbalance of brightness, what is needed is not necessarily an even level of illumination throughout the room but rather some form of compensation for the feeling of deprivation that those sitting at a distance from the windows may have. Windows on more than one side of the room, or in the roof, will help to prevent an excessively steep gradient of illumination. If this is not possible, a great deal can be done by providing surfaces, especially the back walls and the floor, of high reflectance to make as much use as possible of the horizontal flow of light from the distant side windows. This will give a balance of brightness which will ensure that the level to which people in the room will adapt will, in general, be much the same wherever they are whether they are looking down at their work, or looking up and around them. Since as much as half of the total daylight at the back of a side lit room can result from reflected light, the importance of attention to detail in decoration, not neglecting the floor, can immediately be seen.

The absence of glare is essential in good lighting. All view of the sky or of the sun through the window does not have to be obscured, but controlled. There are times when people will want to look through the window at the bright sky and even if they do not wish to look directly at the sun, shafts of sunlight coming through the window can relieve the boredom of the long afternoon. Glare is essentially a matter for skilled control. One of the difficulties, however, is that however skilled the designer may be in providing adjustable controls for use when required, it is unfortunately all too often the case that comparable skill is not used by the people who inhabit the building. Over and over again one goes into schools and hospitals, where staff are educated and intelligent people, and find that the devices which have been provided for the avoidance and control of glare are either not in use, or are misused in a way which seems incredible. The faults are sometimes due to the manufacturer—blinds are sometimes badly designed and fail in service too often, but there is too much unthinking manhandling by preoccupied staff.

Glare can be caused by the presence of windows close to or behind visual focal points, such as a speaker's rostrum or a chalkboard or a swinging rope in the gymnasium (Plate 28). Sometimes this can be avoided by commonsense placing or design, sometimes not. Glare is also caused by large, bright windows on the fringe of the general field of view, so that the eye is attracted and held. Glare is also troublesome due to the reflection of the bright sky or the sun in polished surfaces inside and sometimes outside the room.

To avoid glare, it is important first of all to try to avoid putting windows near to focal points. This is obvious, but is often forgotten. Where it cannot be avoided, the use of blinds or screens over the offending part of the window is essential.

A great deal can be done to reduce glare by attending to detail of the window design. The surrounds to the windows should be lightened both by raising the general brightness of the room with a high proportion of indirect light, and, if possible, by cross lighting from secondary windows in an adjacent wall. The frames and reveals of windows should be designed to give a gentle gradation in brightness from the inside of the room to the sky outside. Frames and glazing bars should be painted in white and the window wall in a light colour. Buildings in traditional brick construction had thick external walls and a vocabulary was developed for grading the light into the room through the medium of splayed reveals. Modern buildings often have relatively thin external walls and special care must therefore be taken to avoid glare, even by providing deliberately a surround to the window to perform the same contrast grading function as the old splayed reveal. Tapered glazing bars and deep sills can all be used to achieve a comfortable gradation of light. The essential is to avoid a hard edge to the boundary between sky and room.

Even when all these details have been attended to, there will still be occasions when glare will be caused by the sky because it is so very bright. Glare is not entirely a matter of contrast, because the eye has an upper limit beyond which it will not work efficiently even if there is no harsh contrast to aggravate the situation. Everyone knows that dark glasses are necessary when out in the snow on a sunny day. This is nothing to do with contrast, it is because the visual mechanism is saturated by the excessively high brightness, and acute discomfort results. When the sky is very bright, the most sensible and appropriate way of controlling glare is to use adjustable venetian or roller blinds, or light draw curtains. These will also serve to protect people from the unwanted heating effects of direct rays of the sun.

USE OF MODEL ANALOGUES

It will be evident that there are always several possible solutions to a daylighting problem. An adequate quantity of light can be provided by windows on one side only, or on two or three sides, by clerestory lighting, by roof lighting, or by combinations of any or all of these. Experience shows, however, that the character of the room and its other less measurable qualities will be different with each solution. At this stage in design the architect has to weigh up a complex series of arguments, some for and some against each solution.

The use of simple scale models as a working analogue will often prove helpful when comparing alternative daylighting proposals. A model made of card to a scale of about $\frac{1}{4}$ inch or $\frac{1}{2}$ inch to 1 foot (depending on the size of the space) will be quite sufficient. It need not necessarily be coloured but the reflection factors of the main surfaces should be reproduced. Viewing will have to be through apertures which should disturb the geometry of the room as little as possible.

If more detailed decisions about the quality and character of a lighting scheme have to be made, or if daylight factors and luminance values have to be checked, then a more accurate model to a larger scale will be required. A scale of 1/12th or even 1/6th full-size is needed to incorporate sufficient detail and to permit a binocular appreciation of the interior. The colour and texture of surfaces should be shown; furniture and other objects in the room should be represented; and it will be found that details such as white painted trim to doors and glazed panels in cupboards are needed to represent the design satisfactorily. Time and trouble have to go into the making of such a model, but it is likely to be of good service to the client, and will permit the solution of many problems at present beyond mathematical engineering.

CHOICE OF GLAZING SYSTEM— TOP LIGHTING AND SIDE LIGHTING

Considering efficiency alone, horizontal glazing placed in the roof immediately above the working plane is the best solution, as not only is the sky brightest at the zenith (that is, the standard overcast sky), but the light reaches the working plane at normal incidence. In factories with roof construction that allows horizontal or low-pitched glazing, it is possible to get an evenly distributed daylight factor of 10% with an area of glazing which is only 30% of the floor area. An equivalent area of glazing in a room lit entirely from one side wall only could give this value of daylight factor only over a small part of the room near the window of the order of the height of the ceiling in from the window. Top lighting is much the most economical method of obtaining a high level of daylight over a large area.

Top lighting is, of course, only possible in single-storey construction or on the top floor of a multi-storey building. Again, although a rooflight transmits light more efficiently than a side window, it is usually more expensive per square foot than a window in a side wall. There are greater difficulties in cleaning rooflights, in the arrangements for sun blinds and blackout, and in providing ventilating opening lights. Buildings with a large roof area and thus extensive foundations tend to be more expensive than more compact buildings which approximate more nearly to a cube. Consequently the economic advantages of top lighting need careful examination in relation to particular requirements. The use of roof lighting as the sole source of natural illumination is usually confined to factories, workshops, and other buildings where deep, continuous spaces have to be lit. Roof lighting has also been used, however, where there are special lighting requirements or where the walls have to be kept free of windows, for example, as in a school gymnasium. Roof lights are also often used in combination with side lights to round out the lighting in a deep room or to focus attention on particular areas. The skilful use of toplights in school classrooms can draw attention to special areas and create a focal point without the need for using tangible screens.

Multi-storey buildings have to be lit by windows in the side walls only unless expensive stepped sections are resorted to. Rooms with unilateral daylighting will tend to have steep gradients of illumination across the room, and there will be glare and unrelieved heavy shadowing which can be somewhat mitigated by the use of surfaces of high reflectance throughout the room. In buildings where there is a particular need for a high degree of visual comfort, such as hospitals, special sectional devices have been used to reduce glare in the areas near the windows and to produce a more even distribution of light across the room. More commonly, however, the drawbacks of unilateral lighting will be relieved by increasing the amount of reflected light through the use of light decorations, particularly on the floor and rear walls. With care it is possible to provide unilateral lighting up to a minimum daylight factor of 2% with comfortable visual conditions in rooms up to 24 ft. deep, and with ceiling of height no greater than 9 ft. To achieve this, however, the closest attention to detailing of the reflectances of interior surfaces is necessary and it is absolutely essential that those who subsequently maintain the building appreciate that the reflectances of the decorations are an integral part of the design.

It is better whenever possible to place windows in more than one wall, so ensuring cross lighting and a greater possibility of working comfortably in any part of the room. Multilateral lighting will give the architect greater freedom in placing windows to meet the users' need for local emphasis or a higher level in particular parts of the room.

Windows which have their heads close to the ceiling will admit more light further into the room in relation to their glazed area than will low windows, as a greater proportion of light will come from the brighter sky closer to the zenith. For the same reason, however, there is a greater risk of glare from such high windows and clearly special care needs to be taken to use blinds and other devices when the sky is excessively bright.

A window or rooflight which meets a wall at right angles will often throw a bright shaft of light across the wall particularly on sunny days. This may be pleasing and stimulating on some occasions, but it can also be distracting if there is some visually important object on the wall such as a chalkboard, a control panel or exhibition pictures. In such cases it is advisable to keep the window reveal at a distance from the wall so that the light flows more evenly along it. Alternatively where careful user control can be relied upon, an adjustable blind should be provided.

OTHER FUNCTIONS OF WINDOWS

The placing and design of windows will also depend upon other considerations such as the desire for privacy, or conversely the display of the contents of an interior of a shop to passers-by. There may be a wish to gain a particularly good view, or conversely to avoid a noisy outlook; to avoid obstructions to light or to secure adequate ventilation. The size or position of windows may be affected by structural limitations of fire regulations. These matters need to be carefully weighed and reconciled with the lighting requirements.

The architect will at the same time be considering the scale and proportions of the windows in relation to the design of the building as a whole, and the sense of enclosure, or alternatively of a link with the outside. The effect of the windows on proportion and scale is vividly illustrated in eighteenth-century houses in which the original small panes and glazing bars have been replaced by single large sheets of glass, giving the façades a blank empty-eyed look. The ability to make large sheets of glass was, of course, a technical advance, but such developments need to be absorbed into the repertoire with a respect for basic principles. The facility we have today to glaze whole walls and to achieve a free flow of space between inside and outside is one which should not be applied indiscriminately. It has to be considered in relation to the possibility of sky glare, of heat losses in winter, and of unwanted solar heat gain especially in summer, in relation to both capital costs (windows are often more expensive than solid walls) and maintenance costs (window cleaning is surely the most expensive way of maintaining the outside of a building).

Attempts have been made to assess the real costs of windows, working out the financial consequences of the various functions and attributes. However, the primary influences on window design are the decisions which the architect has to make about the part the window plays in lighting. It is these considerations to which the architect will return at each stage of his work.

SUNLIGHT

The feeling of well-being which sunlight brings is a welcome amenity, particularly in the winter months, and most people in temperate maritime regions of the world like to have sun in the rooms in which they spend their day whether at home, school, office, or factory, provided that they have it under control. In winter some heat gained from the sun may be welcomed as an economy in fuel, provided that the heating system has sufficiently flexible control to make allowance for the extra heat input. Otherwise the solar heat gain, even in winter, may be a nuisance.

Too much uncontrolled sun will have serious disadvantages because it can cause distress from high temperatures built up by the 'greenhouse effect'. Radiation from the sun penetrates window glass with relatively little loss and then heats up the interior

7

of a room. These interior surfaces then radiate the absorbed heat, but at a longer wavelength than that of the incident solar radiation, and window glass is not transparent to this longer wavelength radiation. As a result, this secondary heat cannot escape and the temperature gradually builds up in the room to a point where it can be uncomfortable to the human occupants. Uncontrolled sun radiation can also cause distress by the direct rays of the sun striking the body.

Excessive glare can be caused by the sun's disc being visible through a window, or by the excessive contrasts produced when direct sunlight falls on a work bench or chalkboard. The disabling glare caused in this way can in some situations be dangerous, and the discomfort from such excessively bright areas is always distracting and often intolerable.

The architect has to arrange windows to admit the sun when and where it is required, and to provide controls to exclude excessive heat and glare. There may also be some spaces in a laboratory or factory, for example, from which all sunlight must definitely be excluded, particularly if the processes require strict control of the internal thermal environment.

Orientation of windows should be considered in relation to the time of day when sunshine is required. The altitude and bearing of the sun at any time in the year and at any time of day can be deduced by one of the many simple methods which are available, for example the Burnett diagrams. For most purposes precise orientation is not important. What is needed is to decide at what time of day sunshine is wanted and then to make sure that some windows, not necessarily the main ones, face in the right direction.

It may also be worth considering the effect of the direction of sunlight on the prospect visible through windows with different orientations. Scenery, whether urban or rural, will normally be more interesting when viewed with sunlight coming from one side than with the sun either directly ahead or behind.

Sun controlling devices include the obvious one of the orientation of the windows. The main windows should be orientated away from the hottest sun, not necessarily the south. South-east or south-west façades receive sun at a lower angle and therefore more nearly normal (perpendicular) to a vertical window and hence the total solar heat intake may be comparable with that for a south facing façade. Control by orientation is particularly valuable in rooms which necessarily have an uncomfortable internal climate, such as kitchens. Methods are available for estimating the solar heat gain through windows and these methods may be used in such critical situations when considering orientation and ventilation requirements.

Sunlight may also be controlled by fixed shading devices or by adjustable blinds. External sunbreaks are far more efficient than internal blinds in reducing unwanted heat gain because they prevent the greenhouse effect. Permanent external sunbreaks such as canopies, louvers or screens, are common in the tropics but they are rarely necessary or suitable for the climate of Britain because they cut down the amount of light entering the room in winter. This is not a serious matter in the tropics, but it is in places with a high frequency of dull overcast skies. In Britain the most appropriate method of control is by an adjustable device (Plates 24, 25, 26, 27). This has to rely for its effectiveness upon the commonsense and intelligence of the people who operate it. It may be better to provide some form of permanent control if staff cannot be relied upon to use an adjustable control sensibly.

Blinds or curtains fixed inside the glass reduce glare and they also cut out direct solar heat radiation, but they are not effective in cutting down unwanted solar heat gain. Once the heat from the sun has entered the room through the window, any surfaces, including the blinds themselves, heat up and the resultant radiation does not escape. Projecting shop blinds, shutters of the traditional patter, and wood lath blinds of the continental type are available, but there is a need for new types of blinds with good weathering properties suitable for modern forms of fenestration.

In order to ensure visual comfort, the construction and material of a blind should be so chosen that the blind is neither so bright as to become a distraction or a source of glare in itself, when the sun shines on it, nor so dark that it reduces unduly the amount of light entering the room, or makes a glaring contrast with the scene outside. It also should be remembered that a blind of dark colour which reflects little light is not appropriate when the blinds are drawn after nightfall. It will not only absorb light and reduce the reflected component of the artificial lighting, but it will also distract the attention because the eye turns not only to the brightest objects in the field of view, but also to things which create an excessive contrast with surroundings or a disjointed effect in an otherwise integrated whole.

For most purposes, suitable materials would be those which diffuse the sunlight and give an internal surface which does not appear excessively bright under the conditions of use when lit from outside by the sun. If the colour chosen for the materials is not neutral or near-neutral, the room will, of course, become suffused with colour. This may be a wanted or an unwanted effect. Window blinds may be used not only to reduce glare from the sun, but they may also be necessary to reduce glare from bright, cloudy skies. Their installation should therefore not necessarily be confined to windows facing the sun.

REMODELLING OLD BUILDINGS

The design of daylighting in a room is not to be confined to new buildings only. It is often possible to transform an existing building which may still have a reasonable expectation of life and make it more suitable for work under modern conditions. Many schools, offices, hospitals, and dwelling houses would gain considerably by alterations and improvements.

In many old buildings such as medieval churches or late eighteenth- or early nineteenth-century houses, the lighting is often of high quality already. Tall Georgian windows with slim tapered glazing bars and white painted splayed reveals are often examples of the highest quality of daylight design and in such cases great care should be taken to preserve and perhaps restore both these good features and the window disposition which may be essential to the proportions and balance of the exterior. Even in many nineteenth-century buildings, many of their positive characteristics will depend upon the means of daylighting and the arrangement of the windows. When architects today wish to make improvements, a nice sense of propriety is called for to decide what should be sacrificed and what should be retained.

It has to be borne in mind, however, that the activities which went on in many old buildings, for which the windows were designed, are entirely different from what may now be required of the building. Elegant old houses transformed into government offices cannot possibly function satisfactorily if left as they are. The visual tasks of modern life are so different from, and often more difficult than those for which the buildings were originally designed that unless careful attention is given to the lighting, people may find themselves working in charming but highly fatiguing surroundings.

Each remodelling job will present its own particular problems and therefore in each case the designer will have to work out the best solution from first principles. Where the lighting in an old building is below modern standards it may be found useful to consider possible improvements in three stages:

(1) Lighter decorations and finishes, to enhance internally reflected light and to assist in buffering glare.
(2) Improvements to existing windows to allow better penetration of light.
(3) More extensive structural alterations.

The redecoration of a room in lighter colours and the use of light coloured finishes, particularly on the floor, will often bring marked improvements to both the quantity and quality of light in the room. Raising the average reflectance will substantially increase the internally reflected component of daylight, while lightening the frames and surrounds to the windows will, together with the increase in internal light, help to reduce sky glare.

If redecoration does not bring sufficient improvement, it may often be found that the window frames, casings, and glazing bars in many old buildings take up a disproportionate amount of the structural opening. Changing the detailing of a window requires care in design because placing a single blank sheet of glass in the window frame of an old building may give an incongruous appearance. Nevertheless by replacing the old windows, which may in any case need renewing because they are falling to pieces, with new windows with lighter frames with tapered bars, it is possible to make a marked increase in the total amount of light entering a building without destroying the character as seen from outside or inside.

Beyond attention to redecoration and to the windows themselves, any further increase in daylighting will require the enlargement of the structural openings of the existing windows, or the insertion of new windows in the wall. New windows should be placed to give the maximum advantages: not only to increase the amount of daylight, but also to give cross lighting if possible, thus lightening shadows and reducing glare from the main windows, or to give a pleasant view out or to allow sunlight to enter a room which previously had an exclusively northerly aspect.

If redecoration is not the complete answer to the daylighting problem, and if changes in the windows are not likely to be successful, or are not possible, or their cost is excessive, then it may be worth while to consider improving the working lighting in all or part of the building by the use of permanent supplementary artificial lighting as described in Chapter 6. Such lighting will make use of fluorescent tubes, and this in itself sometimes causes a shudder of horror in a sensitive architect who is concerned with preserving the entire amenity of the old building under his care. Skilfully designed fluorescent lighting should cause no greater disturbance to the character of an old building than would the letting in of daylight through a supplementary window in the ceiling. A well-designed louvered laylight made of delicate translucent material and with an appropriate scale and details can be sympathetic to an old building.

THE RELATION OF NATURAL AND
ARTIFICIAL LIGHTING

Until recently there was no question but that the main lighting of a building during daylight hours should come from natural sources, the sun and the sky. With the introduction of the fluorescent lamp, however, with a colour approximating to daylight, mixed daylight and artificial light is now common and is not actively disliked. Lighting technology is tending towards the concept of a dual system of lighting comprising the building lighting, which is general throughout the environment, and the work lighting, which is there to highlight and maintain attention and to enable the worker to operate with the greatest visual efficiency. The building lighting can derive equally well from natural or artificial sources or both. Many industrial buildings are now provided with structural roof designs in which both glazing and fluorescent lighting are built in to supply building lighting of the same quality and distribution by night and by day. In such buildings the integration of natural and artificial sources is complete. The proper control of artificial light in relation to available daylight is embodied in the technique of permanent supplementary artificial lighting in interiors (P.S.A.L.I.) discussed in Chapter 6.

Integration of natural and artificial light in this way supposes that the character of a building should be the same by day as by night. This may be true for work places, but in many other buildings such as dwellings, hospitals, schools, and pleasure buildings, the use of the building changes after darkness has fallen either to a more restful, a more stimulating, or a more provocative environment. The lighting should change accordingly.

In a hospital, for example, the lighting by day should be cheerful and bright without being aggressive. Large windows providing a good view to all, but well furnished with blinds which are used to cut out sky and sun glare, are certainly desirable. Light-coloured decorations ensure a bright atmosphere and the absence of dark and gloomy corners. In deep wards, however, some P.S.A.L.I. can be provided to maintain good lighting throughout, while at night a change to a restful character may be desirable. No attempt should be made to emulate after dark either the levels or the distribution of the daylighting. A new character and so a new lighting system should be brought into operation.

In a school, the problem is somewhat different. While here again the building should be bright and radiant by day, conducive to attention and alertness, nevertheless by night there must still be a stimulating atmosphere. Often both staff and pupils work in the evening after a normal day's work, and any soporific tendency in the lighting must be avoided. Here the radiance cannot come from large areas of bright sky as by day, and any attempted imitation would only perpetuate the worst faults of glare. The artificial lighting should by preference then come from sources not bright enough to cause glare, but with sufficient sparkle to give positive pleasure and satisfaction while not causing distraction. Such matters are easy to express in terms of concept, but difficult to design in terms of reality.

Other buildings do not lend themselves to generalisations. Both the daylighting and the artificial lighting in a church, for example, must take tradition into account, but the artificial lighting can well be employed to reveal the form and tracery of the building which may be obscured by day.

In the home individual preferences are paramount, preferences which have to be generalised in the design of hotel or hostel lighting. A factory canteen may be used for special occasions in the evening. During work time an integrated installation of windows and fluorescent lighting may supply the right character; the same canteen with a change to warm filament lighting with ceiling down lights and point source wall lights with sparkle and interest may transform both the room and the social attitudes of those who come back to it to enjoy themselves.

DAYLIGHT CALCULATIONS

There are methods of different degrees of complexity for calculating the interior daylighting in a room. Reference should be made to specialist treatment of the subject of daylighting (see *Daylighting* by Hopkinson, Petherbridge and Longmore, Heinemann, London 1966), since it is not possible to give more than the very briefest outline here of the different methods which can be used.

SIMPLE FORMULAE FOR THE DAYLIGHT FACTOR

It would seem at first sight that there should be no overriding difficulty in devising a simple formula to give an approximation to the daylight level in a room. Clearly the greater the area of glazing, the more light will penetrate the room, and so a direct relationship between the daylight factor and the glazed area (relative to the total floor area of the room) should be possible. This is true for top lighting which is distributed uniformly over the roof, but the situation is more complex for side lighting because the resulting daylight penetration depends upon the distance from the window. It is certainly possible to express the average daylight level in terms of the glazed area, but this is not what the designer usually needs to know. More often, especially if he is working within building regulations, he has to ensure that the daylight level will not fall below a certain minimum level. If the distribution of the daylight has to be taken into account as well as the average level, complications necessarily have to be introduced into any simple formula, and the formula then ceases to be the easy rule-of-thumb that the designer demands. The experienced designer knows the short cuts he can use in the early stages of a design, but these same short cuts in inexperienced hands can lead to serious error if they are used injudiciously.

TOP LIGHTING

The average value of Daylight Factor in a top-lit room is given by the formula:

$$\text{Average Daylight Factor } (\%) = \frac{F. U. A_g}{A_f}$$

where F is a window factor giving the amount of skylight incident on the roof,

U is a coefficient of utilisation, that is, the ratio of the light reaching the reference plane to the light entering the window. A_g is the actual glazed area of the window, i.e. with obstructions due to glazing bars etc. subtracted from the total window opening.

A_f is the area of the reference plane, in practice the floor area.

For roof lighting on a completely unobstructed site, $F = 1$.

The coefficient of utilisation U will depend upon the type of fenestration and upon the interior reflectances of the room surfaces. The table below gives very approximate values of U for typical fenestrations, but strictly it is necessary to have tabulated values for every design of glazing since some forms of patent glazing, for example, introduce louvering and shielding which may reduce the value of U.

TABLE
Value of coefficient of utilisation U for
different roof lighting systems

Type of glazing	Average interior reflectance	
	About 50%	About 20%
Horizontal	0·4	0·3
Shed roof 30° slope	0·4	0·3
Northlight 60° slope	0·25	0·2
Monitor 60° slope	0·25	0·2
Asymmetric monitor ('B.R.S. monitor')	0·2	0·15
Vertical monitor	0·15–0·2	0·1–0·15

(The above table is derived from *Daylighting*: Hopkinson, Petherbridge and Longmore: Heinemann 1966, with acknowledgements.)

Thus for a northlight roof in a factory with rather dirty walls, providing glazing of 30% of the floor area, the daylight factor is:

$$\frac{F \times U \times A_g}{A_f} = 1 \times 0{\cdot}2 \times 30 = 6\% \text{ daylight factor.}$$

With unobstructed horizontal glazing also of 30% of the floor area the daylight factor would be $= 1 \times 0{\cdot}3 \times 30 = 9\%$ daylight factor, while with light and clean decorations this would be increased to $1 \times 0{\cdot}4 \times 30 = 12\%$.

SIDE LIGHTING

Although simple formulae for the average daylight level in a room with side windows have been put forward from time to time, these have very limited usefulness. They are either so inaccurate as to mislead, or insufficiently simple to attract the architect to their use.

Daylight practice in the U.S.A. employs a formula of a type similar to that given above for top lighting, but its use requires the use of extensive tables of utilisation factor for a wide range of window and room dimensions, and for different positions in the room. When these tables are readily to hand, the method is simple and sufficiently accurate for many purposes, but it does not meet the need so often expressed for a simple 'rule-of-thumb' to be used as a first estimate of the effectiveness of a proposed design.

The simplest rule-of-thumb for side lighting is to multiply the required minimum daylight factor by ten, and to take this as a rough measure of the amount of glazing required expressed as a percentage of the floor area. In an unobstructed side-lit room of rectangular proportions up to 5 × 3, with the main glazing in the long wall, and with the window head reaching to a height of not less than one half the shorter side of the room, this rough guide is often fairly close, provided the room surfaces are light, of reflectance about 30%. Thus, to obtain a minimum daylight factor of 2% in such a rectangular room, the glazed area should be:

$$2\% \times 10 = 20\% \text{ of the floor area}$$

Other formulae have been put forward, of greater complexity and of rather greater accuracy. One such formula is given below, for the daylight factor on a horizontal plane in the same plane as the lower edge of the window. This is:

$$\text{Daylight Factor } (\%) = \frac{10 \text{ WH}^2}{D(D^2 + H^2)} + \frac{4GR}{F(1-R)}$$

where W is the width of the window, H is the height of the window above the reference plane, D the perpendicular distance of the reference point from the window wall, G the actual area of glass and F the area of the floor. All these dimensions must be expressed in the same units, i.e. metres and sq. metres, or feet and sq. feet.

The expression on the left is the sky component, and on the right the internally reflected component.

R, expressed as a fraction, is the reflectance of the *walls*. The ceiling is assumed to be of 75% and the floor and furniture combined of 15% reflectance. An exterior obstruction of 20° is assumed.

In order to use the formula, the designer has to have values of the room dimensions to substitute in the formula, but very often at the preliminary stages at which a simple formula is to be of any use, these dimensions are not known. In his early juggling with ideas the designer would prefer to have as few fixed parameters as possible. However, the formula can be pressed into useful service if something more accurate than the simple 'ten times' rule-of-thumb given above is needed. The formula might be used as follows.

A typical problem that faces the designer in the early stages is to determine the minimum ceiling height or the maximum depth of room which will permit him to achieve a statutory level of minimum daylight factor. Suppose that the room size is known even if only approximately but the problem is to determine the minimum ceiling height. Let the room be a classroom of length 24 ft. and depth 18 ft. The daylight factor must not be less than 2% at a point 15 ft. from the window at 3 ft. above floor level.

Applying first the simple rule-of-thumb, the glazed area should be not less than $10 \times 2\%$ of the floor area, i.e. 20% of 24×18 sq. ft., which is 86·4 sq. ft. Suppose the designer is being asked if he can meet the daylight needs with a minimum ceiling height of 8 ft. The area of the window wall above the 3 ft. level will then be $24 \times 5 = 120$ sq. ft. With allowances for glazing bars, frames, etc. it is just possible to glaze the window wall fully above the 3 ft. level to obtain 86·4 sq. ft. of unobstructed glass. The calculation can therefore go ahead. Taking into account an additional 10 sq. ft. of glass below the reference plane level, which might be in glazed doors giving access to the playground, which will add to the internally reflected component though not to the direct sky component, let the total glazed area be assumed to be 96 sq. ft., i.e. one half the area of the window wall. Then in the formula, $H = 5$ (the window head is 4 ft. above the reference plane, 8 ft. above floor level), $D = 15$, $W = 24$ if it is assumed that the window will extend the full width of the side wall, $G = 96$, $F = 432$, and it remains to know R, the reflectance of the wall surfaces. This may be under the designer's control, or it might already be determined by other considerations such as standard decorations and furnishings. For the moment let an average value of 40%, that is, 0·4, be assumed as being sufficiently close to the eventual conditions in a light-decorated classroom. The formula then becomes:

$$\text{Daylight Factor} = \frac{10 \times 24 \times 5^2}{15(15^2 + 5^2)} + \frac{4 \times 96 \times 0\cdot4}{432(1-0\cdot4)}$$
$$= 1\cdot6 \qquad\qquad + 0\cdot6$$
$$= 2\cdot2$$

Since the total of 1·6, the direct sky component, and 0·6, the internally reflected component, exceeds the required minimum of

2·0 by the small margin of 0·2, the daylight needs can just be met by a ceiling height of 8 ft. The margin is small, but sufficient for a policy decision to be made at this early stage of the design to go ahead with a more detailed exercise on the basis of a ceiling height of 8 ft.

If the daylight factor indicated by the formula had been less than the minimum value of 2%, the procedure would have been to increase the ceiling height and the area of glazing by trial and error. Another possibility might be to try the effect of an increase in the average reflectance of the walls; in the given example if the walls were decorated in very light colour to a reflectance of 70% the internally reflected component would

have been $\dfrac{48 \times 8 \times 0{\cdot}7}{432 \times 0{\cdot}3}$

$= 2{\cdot}1$

indicating that the internally reflected component alone would meet the daylight requirements if such very light surfaces could be provided and maintained in condition. Similarly an increase of wall reflectance to 50% would raise the internally reflected component from 0·6 to 0·9.

These simple formulae and rules of thumb serve a useful purpose only if their limitations are understood. It is because they have been regarded in the past as a substitute for knowledge, and consequently misused, that there has been a reluctance to recommend their use.

Where greater accuracy is necessary, and it usually is, in the design of daylighting, other methods must be used. Most of the methods at present available require the separate calculation of the separate components of the total daylight factor. The procedure is as follows:

(1) Determine the sky component with the aid of a suitable geometric or tabular method: for example, using the B.R.S. daylight factor protractors (see below).

(2) Make an approximate estimate of the magnitude of the

externally reflected component and if this is likely to be significant (this can always be estimated by inspection), use the
B.R.S. daylight factor protractors to determine what would
have been the sky component of the area of the obstruction,
had the obstruction not been present, and multiply this
value of sky component by the relative brightness (luminance) of the building to that of the sky. Brick and stone
buildings with glazing usually have one tenth of the brightness of the sky. A light painted building will be brighter
than this.

(3) Determine the internally reflected component by a suitable
method, such as the B.R.S. inter-reflectance table, or the
B.R.S. inter-reflection nomograms.

(4) Sum the three components (DF = SC + ERC + IRC).

The following is a very brief description of two methods of
daylight factor calculation which are at present in general use.

THE B.R.S. SKY COMPONENT PROTRACTORS

The B.R.S. sky component protractors can be obtained from
Her Majesty's Stationery Office in London and carry a full book-
let of instructions which should be consulted. These protractors
were designed by the Building Research Station to enable the
direct sky component to be obtained from architects' working
drawings. The protractors come in a series of several pairs, the
first of each pair giving the sky component, which would result
from a window of infinite length, and the second giving a correction factor for the actual length of the window. The second series
(1968) of the protractors incorporates both primary—infinite
length—and auxiliary—length correction—protractors in one
circular instrument (Plate 29). Each pair of protractors applies
to a particular slope of glazing with a further pair provided for
unglazed opening.

For example, using the pair of protractors designed for vertical glazed windows, the procedure is as follows (Fig. 12). On

the architect's working drawing, the reference point is marked on the section (the scale of the drawing is immaterial) and sight lines are drawn from this reference point to the limit of the visible sky as seen through the window from this point, taking into account not only external obstructions, but also projecting ledges inside and outside the window. The primary protractor is then laid with its centre on the reference point and its zero line along the horizontal, and the intercept of the sight lines for the upper and the lower edges of the visible patch of sky are read off. The difference between the two values gives the sky component for an infinite length of window. For example, if the upper sight line intercepts at 3·5, and the lower sight line, possibly conditioned by an external obstruction, at 0·3, then the difference between these two values, 3·5–0·3 = 3·2, gives the sky component for an infinite length of window. This value must now be corrected for the true length of the window.

The next step is to find the average angle of altitude of the visible patch of sky. This is simply read off on the angle scale and in the above example is 19·5°, say 20°.

Sight lines are now drawn on the plan from the reference point, once again making sure that they refer to the limits of the visible patch of sky taking into account any external or internal obstructions. The intercept of these sight lines is found on the auxiliary protractor using a semi-circular scale which corresponds to the average angle of altitude of the window. When using the auxiliary protractor, its zero line must be parallel to the glazing in the window. Since a semi-circular scale for a value of 20° is not drawn on the protractor, it has to be estimated by eye by interpolation between the scale for 0° and 30°. This is quite easily done and it will be found that the lefthand intercept may be, for example, 0·3 and the right-hand intercept 0·15. The sum of these two values, 0·45, gives the correction factor for the actual length of the window.

Thus the sky component for this window is given by the product of the reading of the primary protractor and the correction of the auxiliary protractor; in this case 3·5 × 0·45 = 1·44%.

Plate 20. Games Hall at the Harris College, Preston. This hall is used mainly for social gatherings, and for recreational physical activities. The curtain is drawn back for games to reveal a brick wall of medium reflection factor. The saw tooth roof section minimises glare from the rooflights, and a low window gives a view out into a courtyard (left). The average daylight factor is about 3%. The light floor gives a high proportion of reflected light. (Photograph Crown Copyright Reserved.)

Plate 21. The working illumination in these single-storey laboratories at York University is provided by the rooflights, with an average daylight factor of 2%. Side windows give a view out and a horizontal component of light. The distribution of artificial lighting follows that of the rooflighting. (Architects: Robert Matthew, Johnson-Marshall & Partners. Photograph by Keith Gibson.)

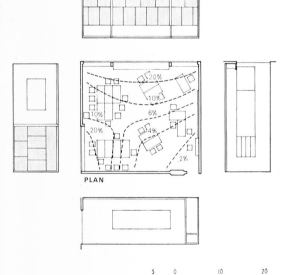

PLAN

5 0   10   20

Plate 22 (a and b). St. Crispin Secondary School, Wokingham. I this room the daylight from the mai window is supplemented by a secon window in an adjacent wall (behin the position from which the photo graph was taken). This rounds ou the lighting and allows the pupils t work in many different positions i the room without shadowing the own work. With a glazed area equa to about one-third of the floor are and a 10 ft. high ceiling, there is high level of illumination, onl falling to 2% D.F. by the doo White venetian blinds are fitted o all windows. The average reflectio factor of the walls is 35%, of th floor 15% and the ceiling 70%. Tw chalkboard fittings give added illu mination on the chalkboard whe required. (By courtesy of H.M.S.O

Plate 23 (a and b). Woodside Junior School, Amersham. Windows placed on three sides of this room, with a roof light placed over the main chalkboard/display wall give even, well-rounded lighting, allowing the children to work easily in any part of the room. Although the glazing area is about 24% of the floor, there is no sky glare as the main window looks on to a shaded court and timber screens are placed to restrict a direct view of the sky. Roller blinds in a light grey woven plastic are provided on the windows to allow control of very bright skies or strong sun. White, splayed reveals to the windows help to grade light into the room. The colour scheme in the room both helps the distribution of brightness in the room and provides a sympathetic background for colourful displays of the children's own work. The average reflection factor of the floor is 35%, the walls 40% and the ceiling 65%. (By courtesy of H.M.S.O.)

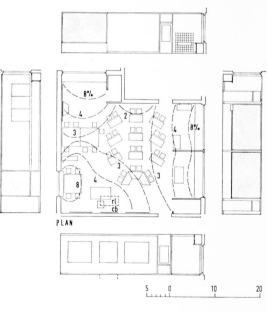

PLAN

Plate 24. Light draw curtains are the simplest way of controlling glare from the sun or bright skies. Such curtains can also be used conveniently to cover the dark windows at night. Suitable materials would be those with a transmission factor of about 40%. (Photograph by John R. Pantlin.)

Plate 25. The spring-roller blinds in this example use an open-weave plastic cloth (Tygan). The fabric breaks the direct rays of the sun but is open enough to allow a measure of ventilation. It permits a better view of the outside than the photograph suggests. (Photograph by John R. Pantlin.)

Plate 26. This external awning keeps the sun off the glass but is high enough to allow one to walk underneath and for doors and windows to open. It also gives a clear view out. It is of course more expensive than curtains or roller blinds. (Photograph by John R. Pantlin.)

Plates 24, 25 and 26 are of the Woodside School, Amersham (see also Plate 23).

Plate 27. This is an interesting example of the most effective means of controlling solar heat again—mounting the blind outside the glass. Tygan cloth which is rot-proof is used, with an aluminium bottom spine running in aluminium channel guides. The open weave of the cloth helps to reduce ballooning in the wind. The blind is operated by a pull cord from inside. St. Peter's School, Gloucester. (Architects: Peter Falconer and Partners.)

Plate 28. Unshielded windows visible to the users of a room which are close to focal points such as a chalkboard or speaker's rostrum can cause glare.

Plate 29. B.R.S. Sky Component Protractor for Vertical Glazing (C.I.E. overcast sky). The Second Series of B.R.S. daylight protractors are available for both uniform and C.I.E. overcast skies. (Reproduced by courtesy of H.M.S.O.)

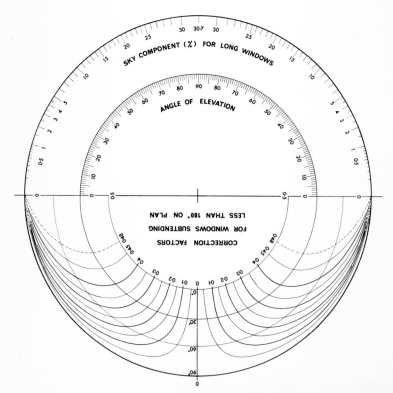

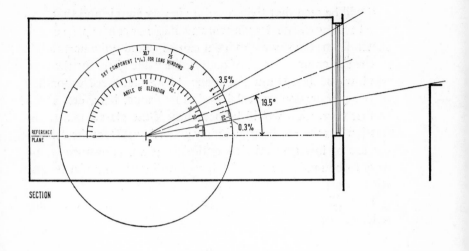

SECTION

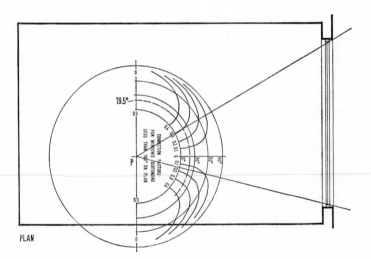

PLAN

Fig. 12. Use of B.R.S. Sky Component Protractors (Second Series)
   Section Sky Component at P for Infinitely Long Window allowing for external obstruction = 3·5%—0·3% = 3·2%
   Plan Correction for length = 0·3 + 0·15 = 0·45
      Sky Component at P = 3·2 × 0·45 = 1·44%

It will be seen that the procedure is very simple and that provided accurate scale drawings are available or can be prepared in
sketch form, the sky component can be found quite accurately.

The sky component is, of course, the value given for the total
window opening. It must be reduced by a factor corresponding
to the obscuration caused by window frame, bars, etc. and a
further correction would be necessary if the glass has a transmission less than ordinary window glass. Apart from the deliberate use of low transmission anti-glare glass, allowance would
have to be made for the dirtying of ordinary glass if a regular
window cleaning maintenance is not available.

To obtain the total daylight factor at this point, the internally
reflected component must be added.

INTERNALLY REFLECTED COMPONENT, THE B.R.S. TABLES

The internally reflected component is found most easily by the
use of the B.R.S. daylight tables. (*Simplified Daylight Tables*,
H.M.S.O., London, 1958, reproduced in *Daylighting*: Hopkinson, Petherbridge and Longmore; Heinemann 1966.)

The IRC is tabulated in terms of the proportion of window
glazed area (true area) to the floor area, and of the reflectance of
the ceiling, floor, and walls. A simplified abstract from the
complete Tables is shown opposite.

The assumption is made in this simplified abstract from the
complete Table that the ceiling and floor (with furniture) have
reflectances of 70% and 20% respectively.

In order to use the table, the amount of glazing in the room
relative to the floor area is found by simple arithmetic. This
should include any glazing below the reference plane, because
although this low glazing will have no effect on the sky component, it nevertheless contributes to the internally reflected light.
the average wall reflectance is found under the column for the
appropriate floor reflectance and the IRC is read off in the row
appropriate to the ratio of glass area to floor area.

TABLE

Internally reflected component of Daylight Factor
for side-lit rooms

Ratio of actual glass to floor area (*per cent*)	Internally Reflected Component (%)	
	Light Walls (R = 60%)	*Dark Walls* (R = 20%)
5	0·3	0·1
10	0·6	0·2
15	0·8	0·25
20	1·1	0·3
25	1·3	0·4
30	1·5	0·5
40	2·0	0·6
50	2·3	0·8

These tabulations are taken, with acknowledgements, from *Simplified Daylight Tables*: Hopkinson, Longmore and Murray Graham; H.M.S.O., London, 1958.

In order to make the table simple, only a limited number of conditions are included, but interpolation in the table can be made by simple arithmetic. For example, if the ratio of glass to floor area in the room shown in Fig. 12 is 1:4, that is, 25%, and the average wall reflectance is 50% (i.e. 0·5) the procedure would be as follows:

Under the appropriate columns find that the IRC for average wall reflectances of 20% (0·2) and 60% (0·6) are 0·4 and 1·3 respectively. For a wall reflectance of 50% (0·5) this will give by simple arithmetic:

$$\frac{IRC-0\cdot4}{0\cdot5-0\cdot2} = \frac{1\cdot3-0\cdot4}{0\cdot6-0\cdot2}$$

i.e. IRC = 1·1 (approx.) for a wall reflectance of 50%.

Thus the total daylight factor in Fig. 12 would be 1·4 + 1·1 = 2·5%.

The complete B.R.S. table (not the simplified version shown above) gives an estimated value of the internally reflected component in rectangular side-lit rooms to a sufficient degree of accuracy for most general purposes. Where more precision is required, other methods must be used, among which the B.R.S. Nomograms for Internally Reflected Light and the B.R.S. Split-Flux Formula for the Interreflected Components are probably the best compromise between precision and simplicity in working. Reference should be made to standard textbooks for a description of their use.

MAINTENANCE, DEPRECIATION AND DETERIORATION

Any precision calculation of daylight factor must take into account the effect of dirt on windows, deterioration of decorations of the interior, and the introduction of internal obstructions. These allowances require care in their calculation because indiscriminate application of 'factors of safety' can lead to excessive glazing which may not only be unwanted but which may introduce a definite disadvantage because of sky glare, etc.

Dirty glass required a correction to be applied to the total daylight factor because it affects not only the sky component and the externally reflected component, but the internally reflected component as well. The allowance to be made will depend upon the degree of pollution in the area and the frequency of window cleaning. It should be borne in mind, however, that glass can look very dirty even when its transmission to light is hardly affected. Dirt deposition has a greater effect on horizontal glazing than on vertical glazing because vertical glazing tends to be rain-washed to an extent which keeps the glass transmission value quite high even though the appearance may be unsightly. Horizontal glazing in the roof, on the other hand, is often in an inaccessible place and may be cleaned only once or twice a year. In a dirty industrial area the light transmission in a factory roof may easily drop to less than one half of the initial value. Even so, however, the decision as to whether or not the glazing area should be increased to allow for this depreciation requires careful thought, because the cost of internal heat loss, etc., might well be greater than the cost of additional permanent supplementary artificial lighting.

Window bars and frames must be allowed for in all daylight calculations. The B.R.S. protractors, for example, are usually set along sight lines corresponding to the total aperture in the wall. If this is done, then the total daylight factor (not only the sky component) must be reduced by the ratio of the actual glass area to the area of the window opening. This correction factor is usually of the order of 0·75. On the other hand, with heavy wooden glazing bars and frames the factor may be 0·5 or less.

Some types of glazing, such as strengthened glass for use in roof lights, resin bonded plastic material, or selective heat absorbing glass, and laminated insulating and heat absorbing glass have transmission factors considerably less than that of clear glass. Selective heat absorbing glass, for example, has a transmission factor of about 65% of that of clear glass, and hence the sky component as calculated with the aid of the protractors or the B.R.S. tables will only be 65% of the indicated value. The weathering of plastic roofing materials can seriously reduce the transmission to light.

Internal decorations can depreciate and so reduce the internally reflected component. In industrial buildings with poor maintenance, the depreciation of the decoration due to dirtying may be so considerable that it may be better to ignore the internally reflected component of the daylight factor entirely. On the other hand, in a non-industrial building or dwelling with central heating, in a clean area, and with regular redecoration every three years, the deterioration in the internally reflected light is so small as to be negligible.

Any depreciation in the IRC due to internal dirt is, of course, in addition to the reduction due to the reduced transmission of the glazing.

DAYLIGHT MEASUREMENT

Daylight photometry should not be undertaken by the inexperienced because it poses all the regular problems of the measurement of illumination together with two additional aggravations, the need to measure simultaneously two levels of illumination, that outdoors and that indoors, differing by 100–1 or more, and the need to measure light which reaches the receptor at near glancing incidence.

The second of these difficulties is met by a special design of photometer which incorporates a degree of 'obliquity error' correction not normally necessary in the photometry of interior lighting. The first of the special difficulties of daylight photometry is insurmountable. Simultaneous measurement of the outdoor and indoor illumination is strictly necessary in order to obtain the true daylight factor fraction, and can be effected only by the use of two light receptors, one exposed to the totally unobstructed sky and the other to the illumination in the room, the response of the two being evaluated by a suitable electrical network to permit the direct reading of the ratio of the indoor to the outdoor illumination.

Various forms of daylight photometer have been designed to try to overcome these basic difficulties. Reference should be made to a standard textbook on daylighting for a thorough treatment of the problem. It cannot be sufficiently repeated, however, that daylight photometry is not an activity for the unsophisticated. Unskilled attempts can only bring the work into disrepute.

Nevertheless the architect and environmental designer often needs some approximate indication of the daylighting achieved in the interior of a building, even though he may not be so rash as to quote the figure which he obtains to anyone but himself and his immediate colleagues. The Building Research Station have designed a simple daylight factor meter which gives a reading of the same order of accuracy as the commonly used photographic 'light meter' and which is simple in operation. The instrument consists of a single photocell which is used for measuring both the outdoor and indoor illumination. In operation the meter is taken to the open window and a sky mask is dropped over the photocell which limits the amount of light reaching it and also confines the direction of the incident skylight to a narrow angle at which the fully overcast sky has a brightness (luminance) numerically equal to the illumination from the whole unobstructed sky. With the meter thus in position at the window, a sensitivity control is adjusted to set the meter indicator at a calibration mark. The meter is then brought to the reference point in the room, the sky mask swung back, and the reading of the meter then gives directly the daylight factor at the point.

The B.R.S. daylight factor meter can only be used when the sky conditions are stable since there is inevitably some interval between setting the instrument at the window and making the measurement in the room.

In bad weather the instrument can be set behind the window glass provided care is taken to ensure that the view of sky through the meter mask is unobstructed by the window bars and frame. The sensitivity control is set as for an open window but the readings of internal daylight factor are multiplied by 0·9 to allow for the effect of the transmission of the window glass.

The B.R.S. daylight factor meter is intended to give a compromise between convenience and accuracy and as such serves a useful purpose. In the event of any higher degree of accuracy being required, it is essential that the professional services of a photometric consultant should be sought.

5

Artificial Lighting

The techniques of artificial lighting, or illuminating engineering, are undergoing radical change and reappraisal in terms of design methods and design principles. The development of artificial lighting during the last fifty years has been essentially a change from local lighting from small light sources of low power to general lighting of the whole building supplemented by special lighting where needed. There has been a halfway stage in which a uniform spread of shadowless lighting over the whole working area has been advocated as the ideal.

The change from small areas of local lighting to general building lighting has been made possible by a steady reduction in the true cost of electrical power for lighting and by a dramatic increase in the luminous efficiency of light sources brought about by the development of the fluorescent lamp. It has been claimed that a lumen of light cost nearly one hundred times as much in 1906 as it did sixty years later. It has also been calculated, probably with less accuracy, that to light the senior common room of a certain Cambridge college with the traditional wax candles costs upwards of two thousand times the cost of modern fluorescent lighting, lumen for lumen.

Most of the technical developments of the last fifty years have been focused—for understandable commercial reasons—on the production of lamps and light fittings with the highest possible efficiency in terms of lumens given out per watt put in. While high efficiency may be welcome in that it means lower running costs, too often it has been achieved only at the cost of lamps with poor colour rendering and fittings that are too bright and cause glare. Light fittings which sacrifice output for appearance are sometimes still dubbed 'architectural' fittings, but fortunately 'Bankers' Georgian' fluorescent fittings are not as common today as they were twenty years ago. The best light fittings today are fine examples of industrial design, approved by both architect and engineer. Nevertheless the best fittings for general lighting are almost always the inconspicuous ones, and many of the most successful lighting schemes are those in which the light sources are incorporated in the fabric of the building rather than merely stuck on the ceiling. To achieve this unified approach to lighting design, architect and lighting engineer have to work closely together from an early stage in the design of the building.

At the same time it is being realised that the most satisfactory lighting schemes, even for work places, are not necessarily those with the highest levels of illumination. Human needs in terms of visual comfort and glare control are now well understood. Some Government Departments such as Education and Health have incorporated these requirements into their official recommendations and standards, and this has helped to raise the quality of lighting design. Although there have been advances in recent years toward giving quality the right priority in lighting design, much has still to be done in order to bring the general standard up to that of the best.

One of the main advances conducive to the improvement of quality of artificial lighting in buildings has been the numerical specification of glare by the Illuminating Engineering Society of Great Britain (the I.E.S. Glare Index System). This system enables a calculation to be made from the known photometric and geometric properties of a lighting installation which will predict whether or not the installation will be within or beyond the limit of glare discomfort considered appropriate for the given building type.

Progress has also been made in the derivation of an index by which the ability of different light sources, especially fluorescent tubes, to render colours can be graded according to the spectral radiation distribution. There has also been progress in the design of artificial lighting to a prescription of the final appearance, through the medium of a scale of subjective brightness and a knowledge of the distribution of light as it governs modelling, emphasis, and the revelation of form.

The simple and basic need of artificial lighting is to 'provide illumination to read fine print'. It has been established beyond any reasonable doubt that the provision of more light enables people to do finer visual work and to appreciate colour in greater richness. This improvement of vision with lighting level is not limitless, however, and although purely for economic reasons the limits have not yet been reached in the general design of artificial lighting, they are now being approached. Summarising, artificial lighting is required:

(a) to light the building after dark;
(b) to supplement daylight where necessary;
(c) to provide special lighting on difficult visual tasks;
(d) to maintain attention on the work;
(e) to ensure safety and alertness.

The first two requirements demand a general distribution of light about the building in sufficient quantity. The last three requirements can be met only by a combination of suitable building lighting and localised lighting controlled strictly with reference to the points where it is wanted. The last requirement also demands the provision of an emergency lighting system to be brought into action should the main lighting fail.

THE BUILDING LIGHTING

Levels of general building lighting have shown a steady trend upwards over the years until the levels provided by artificial lighting are approaching if not exceeding those which result from good interior daylighting. It is probable that a halt in the upward trend of lighting levels can be expected unless the environmental habits of the North Americans become universal, with the consequent isolation of the natural environment and the total dependence upon man-made lighting.

The Code of Recommended Practice for good interior lighting issued by the Illuminating Engineering Society recommends a minimum level of 200 lux for the 'amenity lighting' for totally artificially lit interiors; in addition there is the need for levels of artificial lighting to supplement available daylight of the order of 300–600 lux (30–60 lm/ft²), the exact amount depending upon the available daylight in the room (see Chapter 6). It is a reasonable assumption that for many years to come buildings will be designed with building lighting, that is, general lighting, with average levels of this order both for use by night and in suitably modified form as a permanent supplement to the available daylight.

The distribution of the building lighting has until quite recently had complete uniformity as its objective. Such lighting consists of regularly spaced rows of lighting fittings integrated with the ceiling or suspended from it, the spacing being sufficiently close in relation to the mounting height above the working plane so that the level of illumination over the working place is substantially uniform. Such lighting is so familiar that it is installed without question in working interiors, even though it is not as satisfactory as lateral daylighting in respect of the modelling of faces or the revelation of texture and building form. In fact, opportunities are legion for the improvement of the building lighting in relation to its distribution rather than in any major increases in amount. These problems are actively under study among progressive lighting firms and in the research units of the leading schools of building.

Before leaving this subject of the building lighting, attention should be drawn to the need to select with some care the colour rendering properties of the building lighting not only for its effect upon the colour of people and of objects in the building, but also for its effect upon the building materials. Architects often choose to expose natural and manufactured materials—bricks, 'raw' concrete, natural woods, synthetic plastics, and other building materials—without further applied colours. The great advantage of this, in addition to its aesthetic value, is that the reflectance properties of the building are designed by the architect himself and so he has much closer control of the lighting system in perpetuity than if he had to rely, as is usual, upon the maintenance organisation to understand the importance of the interior decoration in the whole lighting scheme. But if he does choose the materials in this way, he must also select the lighting. Building materials are in general highly sensitive to the colour distortions introduced by many forms of fluorescent lighting—this is particularly true of brickwork and natural timber. A wood which appears warm and restful by daylight can be so distorted by the artificial lighting as to acquire a garish yellowish-green suffusion which is aggressively unpleasant. Some good quality bricks depend for their pleasing appearance by day, especially in sunlight, and under filament lighting, upon some selective reflectance in the red region of the spectrum. Under forms of fluorescent light defective in the red, the effect can be quite deadening. Other types of brick respond to orange-yellow light, and it may have been noticed how pleasant such brickwork is out-of-doors when lit at night by yellow sodium discharge street lamps, but how hideous the same brickwork is by the light of mercury discharge lamps.

No precise guidance can be given as to the choice of lamps to be used with any given building material. The 'de luxe' types of fluorescent lamp are often fully satisfactory, but it is far better to test the material under a range of lamps under proper viewing conditions, because the effects are difficult to predict since they depend upon highly selective reflecting properties of the materials and upon the exact nature of the spectral emission characteristics of the fluorescent lamps.

THE LOCAL OR WORK LIGHTING

The lighting to be provided on the work, in addition to the building lighting where this alone cannot provide fully adequate working conditions, must be determined either by individual analysis of the visual tasks in terms of the critical detail and critical contrast inherent in these tasks or more usually by reference to the recommendations made by statutory or advisory bodies. The most comprehensive set of recommendations is that given by the Illuminating Engineering Society in its Code of Recommended Lighting Practice.

The table below gives a selection of generalised visual tasks together with an indication of the recommended level of worklighting in line with current thinking. The exact change from time to time, partly as a result of changes in the accepted standard of living, and partly because new and more efficient light sources make readily attainable levels of lighting which were previously economically unattainable and therefore of no value in a code of *practice*.

TABLE

Illumination levels recommended for tasks of
different visual difficulty

(The values are given in lux, i.e. lumens/m² in accordance with metric practice. The equivalent in lumens/ft² can be obtained by multiplying by 0·0929, or approximately by dividing the lux values by 10. Thus tasks of average difficulty need 400 lux or 40 lm/ft² approx.)

Difficulty of visual task	Typical examples	Recommended level of illumination (lux)
Little or no visual difficulty	Assembly of heavy machinery	200
Tasks of average difficulty	General machining offices, reading	400
Tasks with small detail	Sewing; business machines; drawing offices	600
Prolonged tasks with small detail	Fine assembly; tailoring; silk weaving	900
Tasks of great visual difficulty	Gem cutting; watch making; fine instruments; hosiery mending	Local lighting to requirements, from 1300 to 3000 lux

The above recommendations are based on the I.E.S. Code of Recommended Lighting Practice, 1968 Edition.

These recommended levels of illumination for the work need modification in certain respects. First, if the operation involves movement, more light may be necessary because movement is seen more accurately and with greater certainty the higher the level of illumination. This will, however, depend upon the nature of the movement. The lighting of machines with regular revolutions or reciprocations requires less modification than the lighting of a task such as a moving inspection belt when an isolated difference in a moving visual field constitutes the essential information for which the observer must be on the alert. The working light can often usefully be increased by up to twice the normal amount depending upon the significance of movement in the operation of the task.

It is also known that older people need more light to see well, even though they may not always express a desire for such light. Unwanted light should never be pressed upon elderly people because some suffer from certain visual defects which are aggravated rather than alleviated by excessive lighting. Generally, however, elderly people with otherwise good sight will benefit by increasing the working light by up to three times, but this increase should always be in the form of selective lighting on the work well screened from the observer's eyes.

The colour of the work and its background may also influence the lighting requirements. Tasks which involve difficult colour judgments need more light but it must be light which actually aids the colour judgment. For example, the recognition of yellow stains on photographic prints, or of the presence of early jaundice in a hospital examination room, is very difficult in a moderate level (100 lux) of filament lighting. Even if the illumination is increased one hundred times to 10,000 lux, recognition is still difficult, whereas a simple change to a 'daylight' type of fluorescent lighting without any change in illumination level will render the colour contrast immediately evident.

The direction of the working light is also to some extent related to the quantity especially where surface texture must be revealed. The recognition of a defect in a sheet of cloth or leather may need an illumination of diffuse light of the order of 10,000 lux such as that obtained from full daylight out-of-doors on a bright day, whereas a carefully controlled flat beam of light giving no more than 500 lux but directed at the correct glancing angle relative to the plane of the material and to the observer's direction of viewing may show up the defect more readily.

Summing up, the essential features of the work lighting are:

(1) Tasks of great visual difficulty require levels of illumination which cannot be supplied either economically or without visual discomfort except by the provision of preferential local lighting on this work.

(2) Tasks of average or little visual difficulty can usually be performed in the levels of illumination recommended for the building lighting and require no preferential treatment.

(3) Some tasks may require work light of special colour or directional properties.

(4) Elderly people need more light, and others with subnormal vision may require special working light directed to the visual task and carefully screened from their eyes.

RELATION OF THE WORK LIGHTING TO THE BUILDING LIGHTING

The relation of the work lighting to the building lighting has already been discussed in terms of general principles. Briefly, it is recommended that the visual task should be brighter than the rest of the room and there should be a gradual change of brightness between the task and the surroundings as seen by the worker at his task, taking into account the recommended ratio of 10:3:1 for the brightness of task: immediate surroundings: general surroundings. This ratio is intended solely as a guide, for a strict adherence to it is not necessary where other considerations may apply. Very often the necessary brightness gradation may be achieved simply by the choice of reflecting characteristics. For example, the brightness of white paper can be satisfactorily graded into the surroundings by employing material for the desk top of reflection factor about 30%, in a room with a floor of reflection factor about 20%.

The task of the lighting designer is to conceive the work lighting and the building lighting as part of the whole environment and to link each constituent part into an integrated whole. The general principles given here and in Chapters 1 and 2 provide the guide lines. Success in the application of these principles can only be achieved with the aid of experience derived from an understanding of the basic thinking behind them, rather than by attempts to apply them by rule of thumb methods. Lighting is both a science and also very much an art, and while the rules of science can be generalised for learning and application, the rules of art are more elusive and require a deeper insight and understanding.

LIGHTING, ARCHITECTURE AND CHARACTER

Beyond creating the conditions for efficient and comfortable visual performance, artificial lighting should be designed to help in revealing in three dimensions the visual qualities of a building, the space and solid volumes it contains and the colour and textures of its surfaces. Normally this will be done by helping to make explicit the structure, form, and materials of the building, but there may be times when the designer will want to use lighting deliberately to conceal structure, to dissolve surfaces, to make scale and distance ambiguous. The best architects of the Baroque period showed with what effect this might be done.

In addition, the artificial lighting has an important and subtle part to play within the design as a whole in helping to establish the character of an interior and thus the attitude or mood of its occupants. Their appraisal of a lighting scheme will depend not only on whether it allows them to see easily but also on whether

° DIFFUSER
cification:—A two tone small section extruded polystyrene diffuser, opal sides
clear reeded bottom section. White cemented end plates for single and twin
> versions. Spine employs A3A cover plate.
inting:—chain or rod suspension, close ceiling mounting and continuous runs
ng A3.4654 jointing piece.

Dimensional, Circuit and Lamp Data QS = Quickstart RS = Resonant Start TS = Twinstart

Fitting Catalogue No.	Tubes No.	Length Ft.	Watts	Voltage. Add correct suffix to Cat. no. i.e. 1 for 200/210 V., 4 for 230/240 V.	Circuit	Length Ins.	Width Ins.	Height Ins.	Weight lb.	Fixing Centres Ins.
A3E/A/Q1040	1	4	40		QS	49·3	3·8	5·1	10·5	24
A3E/A/Q2040	2	4	40		QS	50·3	5·3	5·1	14·3	24
A3E/A/R1065	1	5	65		RS	61·2	3·8	5·1	13·3	24
A3E/A/R2065	2	5	65		RS	62·2	5·3	5·1	19·1	24
A3E/A/R1085	1	8	85		RS	95·6	3·8	5·1	18·8	48
A3E/A/T2085	2	8	85		TS	96·6	5·3	5·1	25·8	48
A3E/A/Q1125	1	8	125		QS	95·6	3·8	5·1	22·8	48
A3E/A/Q2125	2	8	125		QS	96·6	5·3	5·1	31·8	48

5 ft. 80w. fittings available to special order.

photometric data for catalogue no. A3E/A/R1065

Light Output Ratio

Up	Down	Total
11%	56%	67%

Shielding Angle —

Glare Index Data

Lum. Area (sq. ins.)	Flux Fraction Up	Down
232	17%	83%

Reflection Factors			Room Index	A	B	C	D	E	F	G	H	I	J
Floor	Ceiling	Wall	Room Ratio	·6	·8	1·0	1·25	1·5	2·0	2·5	3·0	4·0	5·0
10%	75%	50%		·28	·35	·40	·44	·47	·52	·55	·57	·60	·61
		30%		·24	·30	·35	·39	·43	·48	·51	·54	·57	·59
		10%		·21	·27	·31	·36	·39	·45	·48	·51	·54	·57
	50%	50%		·27	·33	·37	·41	·44	·48	·51	·53	·55	·56
		30%		·23	·29	·33	·37	·40	·45	·48	·50	·53	·55
		10%		·20	·26	·30	·34	·38	·42	·46	·48	·51	·53
	30%	30%		·22	·28	·32	·35	·38	·42	·45	·47	·50	·51
		10%		·20	·25	·29	·33	·36	·40	·43	·45	·48	·50
All	0%	0%		·18	·23	·27	·30	·33	·37	·40	·42	·45	·46
0%	All	0%	Ceiling Factor	·09	·10	·10	·10	·11	·11	·11	·11	·11	·11
			B.Z. Nos.	4	4	4	4	4	4	5	5	5	5

U.F. Conversion Factor for 8 ft. 125W.—0·99:   U.F. Conversion Factor for 5 ft. 80W.—0·97
U.F. Conversion Factor for 8 ft. 85W.—1·04   U.F. Conversion Factor for 4 ft. 40W.—1·02

photometric data for catalogue no. A3E/A/R2065

Light Output Ratio

Up	Down	Total
14%	49%	63%

Shielding Angle —

Glare Index Data

Lum. Area (sq. ins.)	Flux Fraction Up	Down
329	22	78

Reflection Factors			Room Index	A	B	C	D	E	F	G	H	I	J
Floor	Ceiling	Wall	Room Ratio	·6	·8	1·0	1·25	1·5	2·0	2·5	3·0	4·0	5·0
10%	75%	50%		·26	·33	·37	·41	·45	·49	·51	·53	·56	·57
		30%		·22	·28	·33	·37	·40	·45	·48	·50	·53	·55
		10%		·19	·25	·29	·34	·37	·42	·45	·48	·51	·53
	50%	50%		·25	·30	·34	·38	·41	·44	·47	·48	·51	·52
		30%		·21	·27	·31	·34	·37	·41	·44	·46	·49	·50
		10%		·19	·24	·28	·32	·35	·39	·42	·44	·47	·49
	30%	30%		·21	·25	·29	·32	·35	·39	·41	·43	·45	·47
		10%		·18	·23	·27	·30	·33	·37	·39	·41	·44	·45
All	0%	0%		·17	·21	·24	·28	·30	·34	·36	·38	·40	·41
0%	All	0%	Ceiling Factor	·12	·12	·13	·13	·13	·13	·14	·14	·14	·14
			B.Z. Nos.	4	4	4	4	4	4	4	4	4	4

U.F. Conversion Factor for 8 ft. 125W.—1·00   U.F. Conversion Factor for 5 ft. 80W.—0·92
U.F. Conversion Factor for 8 ft. 85W.—1·07   U.F. Conversion Factor for 4 ft. 40W.—1·03

Plate 30. Technical information on light fittings. Clear and full presentation of technical information on light fittings in manufacturers' trade literature is important for both architects and lighting engineers. The page reproduced is a good example. (Courtesy of British Lighting Industries Ltd.)

Plate 31. Glare can be caused by bare lamps or by large, excessively bright light fittings seen against backgrounds which are relatively dark. The brightness of the fittings should be reduced, and that of the general environment should be raised. Glare will be kept within acceptable limits if the I.E.S. maximum Glare Index appropriate to the use of the room is observed (see p. 144). (Photograph Crown Copyright Reserved.)

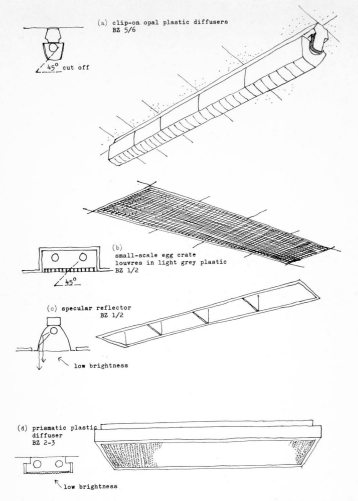

(a) clip-on opal plastic diffusers
BZ 5/6

45° cut off

(b) small-scale egg crate louvres in light grey plastic
BZ 1/2

45°

(c) specular reflector
BZ 1/2

low brightness

(d) prismatic plastic diffuser
BZ 2-3

low brightness

Plate 32 (a, b, c and d). Brightness control of fluorescent fittings. A wide range of fittings is now available to give various light outputs and distributions. Effective brightness control can also be achieved by incorporating the tubes in the fabric of the building. (See Plates 49 and 56.) (*a*) These clip-on plastic louvres were originally designed to meet the requirements of the D.E.S. Regulations, with a 45° cut off and a luminance of about 700 ft-L (7500 asb). They have a light output ratio of 72%. Flicker shields covering the ends of the tubes are desirable. BZ 5/6. (British Lighting Industries Ltd.) (*b*) An unobtrusive recessed fitting designed for use in a P.S.A.L.I. interior (see Plate 38 and Fig. 16). The egg crate louvres have 1 in. by 1 in. cells in a light grey (B.S. 9–094) translucent plastic which have a brightness of about 250 ft-L (2700 asb) and neutralize the blueish appearance of the colour-matching tubes. BZ 1/2. (Elco Plastics Ltd.) (*c*) A metal reflector with an open bottom deep enough to prevent a direct view of the lamps. It has a polished surface and a stepped profile which directs stray light downwards. BZ 1/2. (G.E.C. Ltd.) (*d*) A shallow fitting for surface mounting using a prismatic louvred plastic lens which limits the spread of light sideways. Some light is allowed to spill down the sides of the metal baffle. BZ 3/4. (Rotaflex Ltd.)

Plate 33. Chandelier, New Zealand House, London. In large public buildings today one needs to be able to use lighting to help to create a sense of occasion. This cascade of Perspex prisms down the centre of a staircase is lit from above by flood-lights. It performs the same functions as the crystal chandelier in the 18th century, and presents the same problems—cleaning. (Architects: Robert Matthew, Johnson-Marshall and Partners. Photograph by Henk Snoek.)

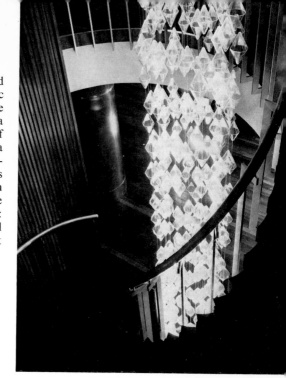

Plate 34. The simplicity both of the form of this lecture theatre and of its lighting and colouring creates an interior free from distractions. The flashing on the upper part of the curtain does not appear as bright as the photograph suggests. (Architect: Denys Lasdun. Photograph by John Donat.)

Plate 35. Students' Union, Durham University. The contribution that a building makes to the urban scene after dark is an important consideration in design. Dunelm House seen at dusk reveals its strong and lively character. It looks even better full of students. The lighting designers were Engineering Design Consultants. (Architects: Architects' Co-Partnership. Photograph by John Donat.)

Plate 36. At higher levels of illumination, glare becomes a more severe problem. In this experimental area at Thorn House, London, the brightness from normal fields of view was kept low by the use of small plastic louvres with a parabolic shape, with a coating having a polished metallic appearance. Their BZ glare classification is BZ1, but annoyance can be caused by the glare from the reflection of the bright tubes above the open louvre cells, if the desk tops, etc., have a highly polished surface. (Photograph by John Maltby, F.I.B.P.)

Plate 37 (a and b). P.S.A.L.I., at the Harris College, Preston. Deep laboratories at the Harris College are lit by a combination of daylight and artificial lighting designed to give comfortable working conditions in any part of the room. The brightness of the upper part of the window is reduced by fixed opaque white louvres, giving a minimum daylight factor of 0·9%. This is supplemented by 5 ft. 80 watt 'daylight' fluorescent tubes installed in a laylight of white plastic cellular louvres, adding about 40 1m/ft² at the back of the room. After dark, the switching is arranged so that some of the tubes above the laylight (those marked N/D) come on with the other fluorescent fittings in the room (marked N). (Diagram by courtesy of H.M.S.O.)

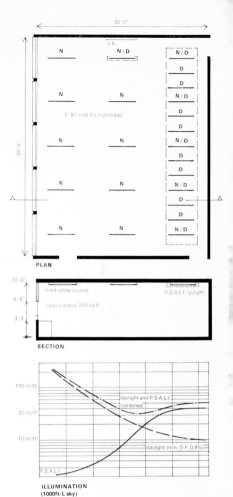

they judge that the character created—lively or restful, neutral or intimate, high key or low key—is the right one. The American architect, Leslie Larson, believes that all lighting should include the qualities of excitement and repose in various combinations. This is perhaps the most elusive of the aims of the lighting designer and is certainly the part of his work least amenable to being subjected to standard routines, yet it should not be neglected merely because it cannot easily be quantified. Fortunately the observation and appraisal of building interiors with successful lighting does suggest certain fruitful lines of approach.

In thinking about the use of artificial lighting to reveal the architecture and to create character, the architect will have considered how the same objectives can be achieved by day. Clearly the artificial lighting design will have to take into account the daylight solution, because many aspects of the room such as the form of the surfaces of the walls and ceiling, the layout and furnishing of the room, and the colour scheme, will have been decided in relation to the design and positioning of the windows. For these reasons the lighting by day and the lighting after dark should be related, although it does not follow from this that the artificial lighting scheme should merely try to imitate the effect of daylight. Two of the chief characteristics of daylight are that it fluctuates throughout the day and that in side-lit buildings it flows across the room from the windows having a marked horizontal modelling effect. Neither of these characteristics is found easily in artificial lighting. Attempts to make artificial lighting fluctuate over a period of time by mechanical means usually appear contrived, however intricate the programme. Variety in artificial lighting rather than change must be sought by more subtle means.

Any attempt to try to imitate the directional characteristics of daylighting by putting the lighting fittings near the windows and directing them into the room, or even by creating substitute or symbolic windows with artificial light sources of large area, will invariably have a very 'stagey' effect apart from creating deliberately the glare problems which good daylight design seeks to eliminate. Artificial lighting can certainly be used to create directional effects but this can best be done by modelling certain objects and revealing certain textures of surfaces within the room rather than by a massive flood of light directed indiscriminately across the room as a whole. Waldram, in his lighting design for Gloucester Cathedral, has shown how the magnificent fabric of the building can be revealed as effectively as in daylight, while at the same time providing adequate light to enable one to read one's hymn-book.

In other more mundane situations directional lighting can sometimes be the main source of both work and building lighting. The most common example of this is probably the domestic interior in which lighting that will allow visual tasks such as reading and sewing to be done easily, and which at the same time helps to give a sympathetic character to the interior, can be created by the use of a number of individual light fittings such as the classic standard or table lamp, which throws a local pool of light on the work, and an indirect flood of light up the walls on to the ceiling.

In most workaday interiors, however, such as large offices, where it should be possible for people to work with equal convenience in any position and facing in any direction, it is usual for the lighting to be generalised with regular arrangements of overhead fittings. This is liable to produce uniform and featureless lighting with unsatisfactory modelling, which can appear to throw an indistinct visual haze over the whole interior. In such rooms where general overall lighting is called for, it is all the more important that the designers, that is, the architect and lighting engineer, should use their skill to give the room interest, definition, and character.

This may be done by the use of suitable contrasts in the choice of surface brightness and colour and by the use of additional directional lighting to reveal texture and form, highlight and shadow on particular surfaces in the room and to create visual focal points. The selective illumination of certain areas of wall, a technique known as 'wall washing', can sometimes serve a useful purpose. An attractive interior can be created by the judicious use of sparkle from metal fittings, mouldings, and small-scale architectural trim with a glossy finish.

Selective preferential local lighting gives scope for a more varied and interesting design. In school classrooms, for example, preferential lighting not only on chalkboards and display surfaces but on local areas where special activities may take place can go far to create the special character of the room. In general offices, laboratories, and drawing offices, additional local lighting for close work not only aids visual efficiency and comfort but can also add to the pleasantness and amenity in the room. In libraries, lights over the tables will aid concentration and maintain the attention, and can also help to give the room the right character. Such local lighting can well be used by daytime as well, with the windows providing the general building lighting.

In buildings with a social and recreational character such as hotels, clubs, restaurants, and theatres, where there are no exacting visual tasks, the designer has greater freedom in the choice of combinations of general and local lighting, of diffuse and directional fittings.

To be successful both in creating the visual conditions which allow people to see well and comfortably and in helping to reveal the architectural qualities of a building, and to create the right character for the interior, it is necessary to ensure

first that the relation between daylighting and artificial lighting is considered from the earliest stages in design, and

second that each is used in a way that exploits its special quality; the variety and directional characteristics of daylighting, the constancy and instant availability of artificial lighting.

THE DESIGN OF ARTIFICIAL LIGHTING
—FLUORESCENT AND FILAMENT LIGHTING

Illuminating engineering is a skilled professional activity. The design of artificial lighting requires a sound technical knowledge of its principles together with an appreciation of the visual results that will be achieved.

The choice of the light source for the lighting of general interiors can be considered simply as the choice between incandescent filament lighting bulbs and fluorescent tubes. In addition to these two commonly used light sources, there are certain types of 'discharge lamp' which have at the moment special application in factories and other industrial areas but which are now under development for more general interior lighting applications.

The relative advantages of filament and fluorescent lighting can be summarised as follows:

Filament lighting is indicated:

(a) where a warm character is demanded;
(b) where running costs are low or are of secondary importance;
(c) where colour rendering is unimportant provided it is familiar;
(d) where simplicity and cheapness in installation are paramount;
(e) where lamps must be changed by unskilled persons;
(f) where bright point source sparkle is desired;
(g) where detailed control over modelling and directional character of the lighting is necessary;
(h) where easily controlled local lighting on the work, e.g. a desk lamp, is indicated.

Fluorescent lighting is indicated:

(a) where running costs over a long period must be minimised;
(b) where high levels of illumination must be provided with the least addition to the total electrical loading in the building;

(c) where uniformity of illumination over large area is needed;
(d) where a cool and hygienic appearance is demanded;
(e) where the lighting is to be used as a supplement to daylight.

Filament lighting is to be avoided:

(a) where the lighting will be used during daylight hours as a permanent supplement and must give a good match with daylight;
(b) where high levels of building lighting will add significantly to the cooling load in the building.

Fluorescent lighting is to be avoided:

(a) where people with abnormal sensitivity to flicker are known to work or live (these are very few);
(b) where colour judgments of special delicacy are to be made (special mixed lighting is then necessary).

The choice of light source need not cause great difficulties. Broadly fluorescent lighting is always indicated for the building lighting, especially where the lighting also has to act as a supplement during daylight hours, and equally filament lighting is always indicated where restricted areas of local lighting under individual control are concerned.

The choice of the precise type of fluorescent lamp is more difficult and is discussed in relation to particular lighting problems at appropriate points elsewhere in the text.

When implementing a lighting design, a decision has to be made either to design the lighting specially for the purpose, for example, to build it into the structure of the building, often a very satisfactory method, or alternatively to make a selection from the very wide range of commercial lighting fittings which are available. The choice must be made in terms of the desired light distribution, that is, the amount of light which is wanted directly on the work, the amount required on the walls, the amount which is required to reach the ceiling, and thence to be reflected back on to the working plane, the degree of freedom from glare which is required, and many other factors which together determine the basic form of light distribution and thence the particular form either of integrated structural lighting or of suspended lighting fittings which the installation will take.

These decisions must be made on the basis of skill and experience, aided where necessary by a full scale mock-up or a carefully detailed trial in a scale model, for while the broad principles of light distribution and its effect on lighting character are known, much more is necessary.

Light distributions are graded broadly into three categories, *direct lighting* which concentrates all the light on the working plane alone, *indirect lighting* which directs all light on to the walls and ceiling to be subsequently reflected back to the working plane, and *general lighting* which distributes light in all directions equally.

It is known, for example, that totally direct lighting on the working plane, while concentrating the attention on the job, is in general unpleasant, gives a 'tunnel effect' because of the resulting dark ceiling, and is considered to lead to visual fatigue because there is no real visual rest centre in the room. Totally direct lighting is therefore in general to be avoided. Where seen, it is often a relic of the days when light was very expensive and had to be directed exclusively to the work and nowhere else.

Indirect lighting, on the other hand, though at first sight pleasant and restful, is not satisfactory for work lighting because the eyes tend to be drawn upwards on to the bright ceiling and so there is difficulty in concentrating attention on the job. Equally it is not satisfactory elsewhere for similar reasons which cause it to lack life and sparkle. Indirect lighting tends to attract the eyes upwards to the bright ceiling, and this is the position which the eyes occupy in sleep, leading to an auto-hypnotic effect, which explains why indirect lighting in a lecture theatre is notoriously soporific.

General lighting is preferred for the building lighting because, provided the design is so arranged that there is no unwanted glare, the overall effect is preferable to most other forms of lighting.

Certain circumstances, however, require a more positive character and this can often be achieved by a combination of general
lighting together with some direct lighting on special areas. For
example, in the chapter on hospital lighting, it is explained how
the lighting of a hospital ward achieves the right character by a
distribution which allows about 60% of the light from the ward
lighting fittings to reach the ceiling, with some direct lighting on
the circulation space between beds, but with no direct lighting on
the patients' eyes. This building lighting is supplemented by
local bedhead lights which patients use for reading, etc.

Indirect lighting dominantly from the walls rather than from
the ceiling sometimes has advantages. These 'wall washing' lighting techniques have a great attraction for interior decorators to
add emphasis in rooms which might otherwise be somewhat
characterless.

LIGHT DISTRIBUTION AND ILLUMINATION LEVEL

The type of light distribution which is given either from the
system of lighting units or from the built-in structural lighting
has a profound effect upon the level of illumination on the working plane. Where the building lighting has to supply the working
illumination in addition to creating the necessary character in
the environment, this influence cannot be neglected. The amount
of light which reaches the working plane from a system of direct
lighting can easily be more than four times that which results
from indirect lighting.

The illumination level on the working plane from a distributed
system of lighting can be calculated very simply. When the
interior lighting is designed around a symmetrical array of lighting units placed in the ceiling or suspended a short distance from
it, the average illumination on the whole of the reference plane is
given by the formula:

$$\text{Average illumination} = \frac{\text{NFUM}}{\text{A}}$$

where N is the number of lighting units;

F is the light output (lumens) of the lamps in each lighting unit;

U is the coefficient of utilisation;

M is a maintenance factor;

A is the area which is illuminated.

The basis of the calculation can be readily understood. If the total light flux available from all the lamps in the installation could be directed to the reference plane and nowhere else, the illumination would clearly be given by the density of light flux in lumens over the area of the reference plane, that is,

$$\text{Illumination} = \frac{\text{Total Flux}}{\text{Area of Reference Plane}}$$

Much of the initial light flux from the lamps is lost in various ways, however. Some light is absorbed in the light fittings themselves because they are not completely transparent or reflecting. More light falls on the ceiling, walls, furniture, and elsewhere and is absorbed and never reaches the reference plane.

The Coefficient of Utilisation, U, handles all these losses of light flux. (The maintenance factor M is simply a factor of safety, usually of the order of 0·8, inserted in the formula to take account of the fact that during the life of the installation, the light output of the lamps will gradually fall, the fittings will gradually become dirty, and the room surfaces will lose some of their reflecting power due to the accretion of dirt.)

The coefficient of utilisation is a means of expressing in one single figure the ratio of the light flux actually reaching the reference plane to the light flux initially emitted by the lamps of the installation. It is governed by (a) the loss of light by absorption at the surfaces of the room, (b) losses due to absorption in the fittings or any other absorption at some stage between lamps themselves and the emission of lighting into the room, and (c) the effect of the proportions of the room. It is usually obtained by reference to tables. The lighting industry, both individual

films and central organisations, issue extensive tables of coefficient of utilisation for a wide range of lighting fittings.

In undertaking a calculation of artificial lighting, it is therefore necessary to know in advance the value of the coefficient of
utilisation which will be applicable to the particular circumstances. In practice the coefficient of utilisation varies from
about 0·15 for totally indirect lighting, through about 0·3 or 0·4
for general lighting, to about 0·5 or higher for direct lighting.

The coefficient of utilisation is usually of the order of 0·4 in the
great majority of lighting installations where the lighting is
deliberately designed to send light flux in all directions and to
reach the reference plane after reflection from the ceiling, walls,
and other room surfaces. The value is, of course, directly dependent upon the reflectances of these surfaces; the figure of
$U = 0·4$ applies to rooms with reasonably light decorations of
overall average reflectance of the order of 40–50%.

Suppose the required average level of illumination in a room to
be 300 lux, a maintenance factor of 0·8 being assumed, and the
coefficient of utilisation having been found from tables to be
0·4, then the total amount of light flux required from the lighting
installation will be:

$$NF = \frac{300A}{(0·4 \times 0·8)}$$
$$= 940A \text{ lumens.}$$

If the total area of the room is 7×10 metres, that is, 70 sq. m.,
a total of $70 \times 940 = 65,800$ lumens would be required.

If twin 4 ft. fluorescent lighting units are to be used for the
installation, employing 'warm white' fluorescent lamps, each of
which emits 2400 lumens, and therefore each pair of which emits
4800 lumens, the total number of lighting units required will be
$\frac{65,800}{4800} = 13·7$, indicating that 14 such twin lamp units would be
necessary. These 14 units would normally be placed symmetrically in the ceiling in order to distribute the illumination as
uniformly as possible over the reference plane.

The calculation is deceptively simple. In practice many things have to be decided and there is a great deal of give and take, if not trial and error, in the use of the method. For example, there may be a good reason, or even nothing more than personal preference, for choosing a particular type of lighting unit. The available wiring in the ceiling may permit only a restricted number of lighting units to be used. Supposing in the same room, there were points available for six fittings only, and the same average illumination of not less than 300 lux was to be provided, the calculation could have been undertaken as follows:

$$6F = \frac{70 \times 300}{0 \cdot 8U}$$

Thus each lighting fitting will have to provide $\dfrac{2625}{U}$ lumens.

If general lighting is to be provided in a room with surfaces of moderately high reflectance, the coefficient of utilisation will be of the order of 0·4. As a first attempt, assume that $U = 0 \cdot 4$. This gives the light flux per lighting fitting to be 6560 lumens.

Reference would now be made to the manufacturer's specification to ensure that the fitting of choice could utilise lamps emitting at least 6560 lumens per fitting. For example, if the preferred fitting employed two 80 watt 5 ft. fluorescent lamps, reference to a lamp manufacturer's tables of lumen output of fluorescent lamps would reveal that a 'natural' type of fluorescent lamp gives a nominal average light output throughout life of 3360 lumens, giving a total of 6720 lumens for the twin tube fitting. This would be satisfactory so far as lighting level is concerned. It would be only necessary then to confirm that the colour of the lamp would be acceptable. If it would not, then the decision has to be made as to whether the desired light output can be obtained sufficiently closely by a lamp of preferred colour or whether preferred colour overrides the precise amount of working illumination (Fig. 14). If it does not, then possibly another type of lighting fitting may have to be accepted.

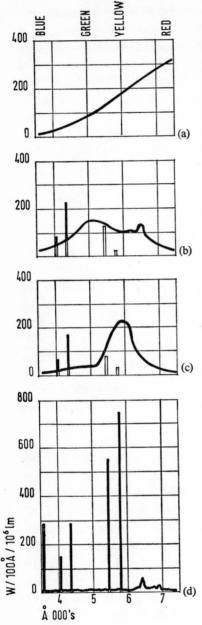

Fig. 14. Colour rendering properties of artificial light sources. Artificial light sources vary widely in the way in which they render the appearance of colours as compared with normal daylight conditions. The colour rendering properties of different sources are linked with their spectral power distribution over the range of visible colours.

(a) The *incandescent* lamp is of relatively low efficiency (150 watt at 13 lm/watt) and has a bias toward the red end of the spectrum. However, it has a smooth curve and acceptable colour rendering properties similar to direct sunlight.

(b) Colour 55 is a *colour-matching* fluorescent lamp with a high colour temperature (approximately 6500 °K), and an efficiency of 44 lm/watt. The appearance of colours including orange-reds and blue-violets, compares well with that under an overcast sky. However, there is still some 'peaking' due to the visible radiation from the mercury vapour discharge itself, and the search for the ideal lamp continues. The appearance of the lamp itself is blueish.

(c) *Warm white fluorescent* lamps are primarily a high efficiency lamp (58 lm/watt) emitting most light in the yellow-green frequences to which the eye is most responsive (see Fig. 1). But blues are weak and reds appear brownish.

(d) *High pressure mercury vapour* lamps are designed for very high efficiency and are not intended to give accurate colour rendering, being very weak on blue-greens and reds. (Based on information supplied by Philips Electrical Ltd.)

Much of the 'give and take' in this type of calculation does not arise in practice. The experienced lighting engineer has 'done it all before' and if he knows his business, he can go fairly directly to the problem. It must also be repeated that this very brief summary can give no more than the broad general principles of the complicated techniques which the skilled lighting engineer will use in designing a lighting installation of a high degree of sophistication.

ILLUMINATION AT A WORKING POINT

The illumination at a specific working point coming from a single point source is given by the formula:

$$\text{Illumination} = \frac{I \cos \theta}{D^2}$$

where I is the intensity of the light source in candelas in the
   direction of the point;
   θ is the angle between the normal at the point and the
   direction of the light source;
   D is the distance of the point from the source.

Such a formula would be used to compute the local lighting at a work place. It would be necessary for the supplier of the local lighting unit to indicate the directional intensity of the unit.

In practice it is often more usual for the supplier to give details of the illumination distribution on a working plane at a given recommended distance between the unit and the work.

Local lighting on machines and other work places which demand concentrated attention should receive illumination which enables the brightness (luminance) of the work to be of the required value in relation to the luminance of the general building interior. For example, if the average illumination in the building interior from the building lighting is 300 lux, and the average reflectance is 30%, the average luminance of the interior will be 100 asb. The luminance of the special task should then be of the order of 1000 asb (with an intermediate grading luminance of 300 asb). If the average reflectance of the task is 10%, then the local illumination on the task should be of the order of 10,000 lux. This would be the order of illumination necessary on a difficult visual task such as sewing on dark cloth.

A long line of linear fluorescent lamps may often be used for the preferential lighting on the work. This would be the case, for example, for preferential lighting on a conveyor belt. A sufficient approximation to the illumination from such a line of fluorescent lamps is given by the formula:

$$\text{Illumination} = \frac{F}{6DL}$$

where F is the light flux in lumens emitted by each lamp;

D and L, both expressed in the same units, are respectively the distance of the reference plane below the line of lamps and the length of each lamp.

In practice again, however, most manufacturers of linear lighting units supply the distribution of illumination on a reference plane for a given mounting height above the plane.

The illumination on a reference plane below a large source such as a laylight mounted immediately above can be obtained from the formula:

$$\text{Illumination} = \frac{L(1-\cos A)}{2}$$

where L is the luminance of the laylight;

A is the angle subtended at the reference point by a circle of the nearest equivalent area to that of the laylight. This angle A can usually be determined by inspection or, if necessary, by calculation.

More precise methods exist for the determination of the illumination below a large laylight, and these methods would be used in the design of lighting for a permanent artificial supplement (P.S.A.L.I.) (see Chapter 6).

LIGHTING QUALITY—DIRECT AND REFLECTED GLARE

Direct glare in artificial lighting in building interiors is almost entirely a matter of discomfort rather than impairment of vision. It is true that in badly lit buildings, the presence of a brilliant unshielded light can disable vision, either by a blinding effect or more likely by causing distraction by phototropism, attracting the eyes which turn instinctively and unavoidably to the bright light. The subsequent effect on vision caused by the after-image of the bright light source may be dangerous. Such occurrences, however, are only likely in buildings so badly lit that the necessary elementary improvements are so obvious that no deep technical knowledge is needed to put them in hand. Where a level of building lighting of the order of that recommended in present-day Codes of practice is installed, direct visual disability due to glare is unlikely to occur.

Discomfort glare, on the other hand, does occur with regrettable frequency when the designer has concentrated more upon meeting a quantity specification than upon achieving overall good lighting. The factors which govern discomfort from glare are well known and are now the subject of precise quantification.

Glare is a function of the brightness (luminance) of the source and of its apparent size, and of these two the source brightness has the greater effect. It is also established that the larger the glare source, the more significant is its brightness so that when the lighting comes from an overall luminous ceiling, the glaring effect is governed very largely by its brightness. The Illuminating Engineering Society recommends that for such an overall luminous ceiling, the upper brightness limit to keep glare below the threshold of discomfort is of the order of 150 ft-L (about 1500 asb).

When the lighting comes from a number of small sources, the permissible luminance limit can be greater (it is set at an average of 1000 ft-L for school classrooms). Glare cannot be controlled universally by a simple limitation on luminance since the size of the glare source is an important factor. Glare caused by the same source is also less troublesome when the source is seen in a bright rather than in a dark environment, and local contrast grading around a glare source can also reduce discomfort. Sources well out of the line of sight are less glaring and so it is obvious that glare is a function of the total environment rather than of the sources alone.

The only complete method of controlling glare is to control the whole environment, and this is the purpose of the I.E.S. Glare Index control system which is based upon the fundamental research at the Building Research Station (see Chapter 3).

The Glare Index is an evaluation in terms of the whole installation, the number of light sources, their brightness, their area as seen by the observer, their position in the field of view, and the brightness and reflectance of the environment in which they are seen. The tables apply to installations of symmetrically arranged sources. The relevant I.E.S. technical publications should be consulted for a full description of the operation of the method. It is not complicated, but it has been designed for operation by the specialist lighting designer.

The Illuminating Engineering Society in its Code of Recommended Lighting Practice gives a recommended value of Limiting Glare Index not to be exceeded in the appropriate environment. Thus in school classrooms, the Limiting Glare Index is 16, in some factories it can be as high as 22 or even 28, while in hospital wards it should not exceed 13. This recognises that glare is more tolerable in some situations than in others. This is, of course, only common knowledge, but it is recognised quantitatively in the I.E.S. recommended values.

While direct glare can now be controlled quantitatively with the aid of the I.E.S. Glare Index system, glare which results from unwanted reflections has not yet been handled quantitatively to the same extent. Reflected glare can cause visual disability, for example, when reflections in the glass cover of an instrument prevent one from reading the dial, and in general it causes dis-

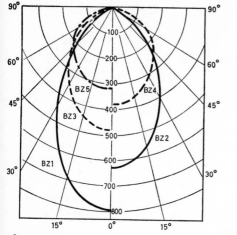

BZ 1	$I \propto \cos^4 \theta$
BZ 2	$I \propto \cos^3 \theta$
BZ 3	$I \propto \cos^2 \theta$
BZ 4	$I \propto \cos^{1 \cdot 5} \theta$
BZ 5	$I \propto \cos \theta$
BZ 6	$I \propto (1 + 2 \cos \theta)$
BZ 7	$I \propto (2 + \cos \theta)$
BZ 8	I constant
BZ 9	$I \propto (1 + \sin \theta)$
BZ 10	$I \propto \sin \theta$

The BZ classification relates to the lower hemisphere only; the polar curves above are scaled to give 1, 000 lumens in the lower hemisphere for purposes of comparison.

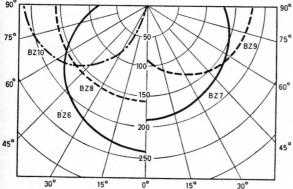

Fig. 15. Polar curves in the British Zonal (BZ) classification of light fittings showing their downward light distribution. It will be seen that BZ1 fittings emit very little light above an angle of 45°, while BZ8 fittings distribute light equally in all directions. (Reproduced with permission from I.E.S. Technical Report No. 10.)

comfort, annoyance, and distraction, for example, when the reflections of bright fittings or even bare lamps are seen in a polished table top alongside the visual task, or in a gloss-painted wall on which a chalkboard is mounted. Uncurtained windows by night are, of course, a potent source of reflected glare, particularly if special glasses of high reflectance are used for the glazing. In order to diminish the risk of reflected glare, care has to be taken in the selection of lighting fittings bearing in mind that a fitting which is carefully designed to eliminate direct glare to people looking in a more or less horizontal direction or slightly upwards may have an open bottom resulting in a reflection of the bare lamp in a polished table top. With single small fittings it is often possible to move one's work or one's head to avoid reflected glare, but with close arrays of fittings, particularly of large linear fluorescent fittings, it is often not possible to avoid trouble in this way. In circumstances where work positions are fixed, as in a drawing office, reflected glare can be a great nuisance.

Clearly the most effective precaution against reflected glare is to prevent the use of any glossy surface in a position where it can cause trouble. Tables and desk tops should have a matt finish and the use of high-gloss paint on large areas such as walls and ceilings should be discouraged. It is particularly undesirable on walls to which chalkboards or pin-up boards are to be fixed. The casings of typewriters and other machines should not have a highly polished finish.

When all these precautions are taken and some reflected glare still remains, it is possible to effect an improvement by introducing a strong component of directional light from an entirely different direction from the main lighting, which will swamp the unwanted reflections. This is a skilled technique and should not be undertaken without proper knowledge.

Many experiments are being undertaken with the use of polarised lighting in the ceiling, sometimes in the form of a luminous ceiling, and claims have been made for the success of such lighting for the elimination of unwanted reflections.

LUMINOUS CEILINGS

A special but brief mention should be made of the technique of using an overall luminous ceiling for artificial lighting. Opinions are divided about the virtues of such a system and the Illuminating Engineering Society has suggested that it is not suitable in general for work places unless supplemented by directional local light, though it may have its place in decorative lighting schemes.

The visible ceiling is a false ceiling, the lamps being mounted on the true ceiling above, with diffusing or louvered material suspended below. In such an arrangement the distance between lamps should not exceed twice the distance between the lamp plane and the plane of the diffusing ceiling or, when translucent louvers are used, the distance between lamps should not exceed 1·5 times the distance of the lamp from the louvers. The cavity in which the lamps are placed must be painted in white of very high reflectance. The efficiency of the system depends upon this inter-reflection of light within the lamp cavity above the diffusing ceiling.

Fluorescent lamps are invariably used in luminous ceilings. Great care has to be taken in the design not to exceed the permissible limits of luminance because otherwise excessive glare will result. The maximum recommended luminance is 150 ft-L (about 1500 asb).

The efficiency of utilisation of luminous ceilings is not high. The chief advantages are in decorative spaces for concealing building services and structural details.

6

Integrated Lighting

The attitude of the worker to the use of artificial lighting during daylight hours has profoundly changed since the introduction of the tubular fluorescent lamp. The poor colour and the stark design of the first lighting units led to a prejudice against this form of lighting which is still not fully resolved for the lighting of social places. For work places, however, the advantages are now conceded of fully adequate illumination on the work throughout the day.

It is very easy to observe, by walking round a city business area during the daytime, that people clearly prefer to work with the artificial lighting switched on during the day. This applies equally to work places with good daylighting as to those which actually need the artificial lighting to enable them to see to get about their business. The whole attitude to artificial light during daylight hours has changed.

It was recognised some time ago, however, long before this change in attitude on the part of workpeople was manifest, that if artificial lighting was to be used successfully during the day, it must be integrated in design with the daylight in such a way that it formed an acceptable supplement to the daylight without introducing a rival dominant characteristic. People who have no longer a prejudice against artificial light still prefer to work in good daylight if they can get it, and the next best thing is to have *apparently* good daylight even though much of the working illumination may be supplied from unobtrusive artificial light sources. This technique of integrating the artificial lighting with the daylighting, in terms of human vision and adaptation, is called permanent supplementary artificial lighting for interiors (P.S.A.L.I.).

PERMANENT SUPPLEMENTARY ARTIFICIAL LIGHTING IN INTERIORS (P.S.A.L.I.)

During the years since the first introduction of the concept of P.S.A.L.I., the change in attitude to artificial lighting at work during the daytime has rendered unnecessary much of the original argument put forward to justify the cost of installation and maintenance of the artificial supplement. Levels of illumination of 300–500 lux (30–50 lumens per sq. ft.) necessary to form an adequate supplement to daylight levels were two to three times the levels which the economy then considered justifiable. Now such levels are commonplace in new buildings, and it is becoming less difficult to have them installed in older buildings.

Nevertheless, although the concept of P.S.A.L.I. is generally accepted, there is a considerable field of study left open for the forms which the design of the installation shall take. The simplest form of lighting during daylight hours is, of course, to provide an installation of artificial lighting, uniform over the whole of the working area, to a level of the same order as that of the average level of daylight in the better lit parts of the room. This installation is used both during the night and during the day. By day, it follows that the areas near the window will receive considerably more than the average illumination, and those remote from the window will receive little more than the artificial light level. Such an installation can hardly be satisfactory to everybody, those remote from the window necessarily being conscious of some deprivation. Nevertheless, such an installation is cheap, because it requires no special design, no special calculations to integrate it with the daylighting, and no special precautions in its use. It is simply switched on, left on and ignored. This not wholly satisfactory solution to the problem of lighting during daylight hours is nevertheless widely accepted in the vast mass of speculative office accommodation in cities and towns all over the world. This passive acceptance has unfortunately been taken by some as the criterion of adequacy.

The true technique of integrated lighting finds its place in buildings where the visual nature of the activity demands care and consideration in its planning. Schools, hospitals, factories for precision working, and drawing offices are typical examples of buildings where carefully planned P.S.A.L.I. is desirable and necessary. In addition, when it is realised that properly planned P.S.A.L.I., though not less expensive to install, may be less expensive to run than the 'packaged deal' system of daytime artificial lighting, it is likely that well designed P.S.A.L.I. can find its place in the best general offices as well.

BASIC PRINCIPLES

In a well designed combined installation, the level of the supplementary lighting is determined by psychophysical considerations of adaptation and of balance of brightness between the better lit areas of the room near the window, and the supplemented areas which receive less daylight. By careful planning, the essential *daylight character* of the room can be preserved. People will have adequate general illumination on their work, which is the first consideration, but in addition the general appearance of the room will be of a room receiving good daylight.

The system offers much greater freedom in design because the proportions of the room are no longer necessarily linked to the requirements of daylight penetration. In addition, by the use of suitable screens on the windows, glare can be reduced because the windows are no longer required to serve as the only illuminant for the working areas of the room. The design of the windows can therefore be related directly to considerations of visual comfort, the need for a view out, the appearance and relation to the form and structure of the building.

The supplementary light has to be distributed about the room in such a way that the brightness of the areas dependent most on supplementary light balances that of the areas near the window. Complete equality is not demanded. It is known, from studies of appearance and apparent brightness, that the visual effect of any area in a room is related not only to the actual physical brightness (luminance) of that area, but is also determined by the general state of adaptation of the observer. This general state of adaptation is influenced by all the areas in the field of view, including the visible sky through the window. Consequently, an area at the back of a side-lit room, for example, although it may receive adequate illumination as measured by a photometer, may nevertheless look dark and gloomy because the physical brightness of the remote areas is well below that of the adaptation level, and therefore the apparent brightness is low. The essential requirement of the permanent supplement, in addition to supplying adequate working illumination, is to bring up the apparent brightness of those areas remote from the window to a level which is acceptably within the comfort range around the adaptation level prevailing in the room as a whole, even though there may be no requirement for additional illumination on the working plane.

The permanent supplement has therefore to be designed not only in terms of the illumination actually falling on the working plane, but also in terms of the distribution of light falling on the walls. The physical brightness (luminance) of these wall surfaces etc. will be determined both by the illumination which falls upon the surface, and by the reflection factor of the surface. While theoretically the same luminance can be obtained by a high level of illumination falling on a surface of low reflectance, or a lower level of illumination falling on a surface of higher reflectance, it is necessary, in P.S.A.L.I. design, to take the effect of *brightness constancy* into account (see Chapter 3). The subjective effect of a high illumination on a surface of low reflectance will not be the same as that of a low illumination on a surface of high reflectance, provided that the eye is able to see the whole visual scene, the sources of light as well as the surface upon which they fall. The surface of lower reflectance will be recognised for what it is. In addition, any surface of higher reflectance introduced subsequently will accentuate properties of the darker surface because of *simultaneous contrast*, i.e. the enhancing of the contrast between continuous areas.

The design of P.S.A.L.I. could, indeed, be a highly complex theoretical exercise. In practice, however, the great majority of working interiors, and many other interiors as well, fall into fairly simple patterns of light, shade, colour, and reflection factor, and good supplementary lighting design is possible from simple concepts of illumination level and light distribution. In addition, it has been possible, by drawing on the results of extensive experiments, to formulate some empirical rules for the design of P.S.A.L.I. which appear, from experience, to give satisfactory results in the great majority of design problems.

DESIGN OF SUPPLEMENTARY LIGHTING

Experimental studies indicated that the level of supplementary light to give a satisfactory balance of brightness with the daylit parts of an interior depends both on the visible sky brightness and on the distribution of daylight in the room. This dependence is direct; the higher the level of sky brightness, the higher the necessary level of supplementary illumination. This is the contrary of what would be expected purely in terms of working illumination. Theoretically, the level of supplementary light should be linked to the available daylight, and should vary with the level of daylight illumination throughout the day, with changing cloud cover, and so on. However, the experimental work demonstrated that the exact level of supplementary lighting is not sensitive, because the adaptation of the eye permits a wide variation about a mean level. If such an average level of supplementary light is chosen, there will be occasions when it appears inadequate because the skylight outdoors is very bright, and there will be more frequent occasions when it appears to swamp the daylight. It has become the practice to design supplementary lighting around an average sky condition giving 10,000 lux (1000 lumens per sq. ft.) on a horizontal plane from the unobstructed sky. Again referring to the experimental background, an empirical formula has been derived which relates the level of supplementary light to the Daylight Factor distribution in the interior. This empirical formula is as follows.

$$E_t = 500D$$

where E_t lux is the average level of supplementary illumination,
$D\%$ is the average Daylight Factor over the area of the room to receive supplementary lighting.

This relation depends on the assumption that the distribution of light from the supplementary sources enables the walls to receive a substantial proportion of illumination, i.e. of the same order as that on the working plane.

The area over which supplementary lighting should be provided will be determined by the distribution of daylight in the room. To a sufficient order of approximation, supplementary lighting should be distributed (a) over the areas of the room in which the Daylight Factor is less than 2% or (b) over the area of the room (in a side-lit room) in which the Daylight Factor is less than one-tenth of the average Daylight Factor distant 5 ft. in from the main window. Strictly, it is necessary to obtain the detailed distribution of Daylight Factor in the room and to provide the supplementary lighting over those areas where the Daylight Factor is less than a given fraction of the average Daylight Factor. In practice the empirical rules, either (a) or (b) given above, will give reasonably satisfactory results. They are not universal rules to be applied without careful thought, however, and care must be taken because it will be obvious on inspection that they cannot apply in some interiors. For example, in a room which receives very little daylight, either because windows are small or because there are obstructions outside, the daylight level even near the window may be quite low, and additional illumination is necessary over a much greater area than that represented by one-tenth of the Daylight Factor 5 ft. from the main window. There may, in fact, be no areas of the room which reach a Daylight Factor of 2%, except very close to the window. Supplementary lighting would, in fact, be necessary over the whole of such a room, and the windows would simply act as apertures to permit a view outside, contributing little or nothing to the working illumination. The essential character of such an installation would be that of a fully artificially lit room, with a view window and a little daylight as a bonus.

The *distribution of supplementary lighting* should be graded from the darker to the lighter parts of the room. The best result is obtained when there is a gradual increase of total illumination up to the window. The level of supplementary lighting should always be less than that from the daylight near the window on average days (i.e. when the sky provides 10,000 lux). Any dip in the illumination level in the middle of the room must be avoided, because it draws attention to the artificiality of the situation.

The gradient of illumination from the side windows to the supplemented areas for unilateral lighting should be such that the illumination over the main working area should not vary by more than 3:1, although the maximum illumination in areas very close to the window can be outside this range. A preferred illumination distribution curve can be specified, from which the supplementary lighting can be evaluated by subtracting the values of available daylight in the interior from the required illumination distribution. In practice, however, it would be very difficult to design a supplementary system exactly to fit the curve so obtained by subtraction. It is better to design the supplementary illumination around an average value of required illumination, say 500 lux, selecting an appropriate light distribution, and thence to determine the illumination distribution curve for the daylight and the supplement combined. Minor adjustments in the design of the supplementary lighting can then be made to ensure that the final curve does not depart too radically from the desired curve. Even more preferable, but possible only in new buildings, or buildings undergoing conversion or rehabilitation, is to design the fenestration and artificial supplement together.

With multilateral lighting, the distribution of daylight will depend upon the placing of the windows, two-side lighting resulting in a dip in the level of daylight illumination in the centre of the room, and adjacent-side lighting resulting in a less well illuminated area in the corner remote from the windows. The same principles apply, but the area to receive supplementary lighting will be different. In the case of two-side lighting the middle of the room will require supplementary illumination. Here again the requirement would be to ensure that, on a day of average sky brightness, the illumination under the supplement would not exceed that near either of the windows. The amount of supplementary lighting should be disposed about the room again to ensure that the diversity in illumination did not exceed 3 : 1, except for the non-working areas close to the windows themselves.

In the case of adjacent-side lighting, the supplement would be required over a relatively restricted area. The distribution of daylight illumination would be required in two directions, with respect to each window, but the general principles of operation would be the same.

DESIGN OF WINDOWS

Integrated daylight and artificial light should be planned from the start, the windows being designed bearing in mind that daylight penetration, with an integrated system, can take second place to the ensuring of freedom from sky glare. Protection from sky glare is obtained (a) by restricting the visible area of sky, (b) by reducing the visible brightness of the sky, and (c) by enhancing the average brightness of the room itself.

The existence of supplementary illumination in the room will raise the average brightness of the room, and thereby of itself tend to mitigate sky glare effects. However, the additional room brightness achieved in this way will not be sufficient completely to buffer the glare from the sky on a bright day. Some additional attention to the window design is desirable. The provision of permanent horizontal louvers, baffles, or other screens can be considered because the reduction in daylight which results from such devices can be replaced by the additional supplementary illumination. Such devices are of considerable assistance in rooms, such as a hospital ward or a school, where the sky glare problem is most serious for occupants near the window. By suitable design of such devices, the amount of visible sky to occupants near the window can be considerably reduced, to the point where, with the additional brightness provided by the supplementary illumination, little glare results except on the brightest days.

Glass of lower transmission than normal, inserted in the upper panes of a window, will reduce the degree of glare for every occupant in the room, and not only for those near the window. The use of such low transmission glasses, combined with integrated supplementary artificial lighting, can give a satisfactory all-round result. Adjustable blinds, either of the venetian type with adjustable slats, or of woven material which permits a view outside, are an obvious solution in situations where the supplementary lighting has to be integrated into the design of an existing building.

CHOICE OF THE ARTIFICIAL LIGHT SOURCE

A permanent artificial lighting supplement must match the available light in colour as well as in brightness, if it is to be unobtrusive and not cause distraction. In addition, the levels of supplementary lighting are sufficiently high to demand for economic reasons the use of a light source of the highest possible luminous efficiency consistent with good colour. Good quality fluorescent lighting is the only light source at present available which meets these requirements.

Fluorescent lamps are, however, made in a variety of colour qualities, very few of which are suitable for matching with daylight. Daylight itself varies in colour quality, from the relatively 'warm' colour of direct sunlight to the very 'cold' quality of the clear blue sky. The diffuse overcast sky has qualities midway between these two extremes.

The colour qualities for an artificial light source to supplement daylight have to be specified in two distinct forms. First, there is the colour of the light source itself, and the appearance of the light reflected from neutral white or grey surfaces. Second, there is the colour rendering quality of the light, i.e. the ability of the light to render different colours in relation to the appearance of these colours in natural daylight. These two specifications are not identical. The eye is not able to sort out the individual colour frequencies in a mixed light source into the individual components, in the way in which the ear can sort out the individual frequencies in a complex chord. As a result of this, two light sources with entirely different spectral compositions, can appear, subjectively, to be of the same colour, and the effect of light from the one source or the other on white or neutral grey surfaces will be the same. On the other hand, the rendering of colours other than neutral will differ, depending upon the component frequencies of the two sources. It is, for example, possible to design a fluorescent lamp whose *colour appearance* is identical with average overcast daylight. Such a fluorescent lamp, however, might be composed of a relatively narrow band of frequencies towards the blue end of the spectrum, another band in the green or yellow-green, and another band in the orange. Red might be completely absent. Such a light source as a permanent supplement to daylight would be completely satisfactory if all the surfaces in the room were a white or neutral grey, and the observer was the sole occupant. The supplementary lighting, judged by its effect on the neutral surfaces in the room, would be exactly the same as the overcast daylight coming through the window.

Once surfaces of different colour from neutral, however, are introduced in the room, the difference between the two light source will be immediately apparent. Many colours will be grossly distorted in the fluorescent lighting which contains no red light. The degree of departure of the colour rendering given by the artificial supplement, as compared with natural daylight, will depend upon the spectral reflection characteristics of the surfaces. Some surface colours may be hardly distorted at all. Others, particularly those like the human skin which depend upon the presence of frequencies in the red range for their correct rendering, will be most seriously distorted.

It therefore follows that satisfactory supplementary lighting will only be obtained provided the colour rendering properties of the supplement are closely comparable with those of the natural daylight in the room. Since natural daylight itself is not constant, a compromise will clearly be necessary. Knowledge of the visual processes which govern colour rendering is still insufficient for a specification to be laid down from theoretical grounds alone. It is necessary to rely upon carefully conducted experimental studies. These have revealed that there is, at the moment, no type of fluorescent lamp which is as good as daylight itself as a permanent supplement. However, there are certain lamps which are preferred above all others for this purpose, and where possible it is these types of lamp which should be used in the design of P.S.A.L.I. If the windows face away from the sun, the best form of light source for the permanent supplement is the type known as 'colour matching', designed by the lamp manufacturers to give a rendering of colour as closely as possible the same as that of daylight without sunshine. If the room is liable to receive direct or mixed sunlight, the 'colour matching' type of lamp, though often giving a fresh appearance, can also appear excessively 'cold'. A better choice of lamp may then be the 'improved daylight' type, which is slightly less bluish, and, although the colour rendering is not as accurate as that of the 'colour matching' type of lamp, nevertheless it is probably satisfactory for all but the most critical situations.

Neither of these types of fluorescent light source is entirely satisfactory, unfortunately, because the 'colour matching' type of lamp is blue in appearance and conveys this bluish hue to white surfaces in the near vicinity. Equally, the 'improved daylight' type of lamp is pinkish in appearance, sufficient to attract the attention in critical situations. An ideal lamp for supplementing daylight will, it is hoped, eventually be produced.

In the meantime, much can be done by the care taken in the design of the supplementary system, avoiding any view of the bare lamps themselves, and also avoiding the use of surface colours in the near vicinity of the lamps (e.g. the louvers or egg-crate diffusers) which draw attention to the colour deficiencies of the light source. This is a matter of experience based upon trial and error with existing supplies from manufacturers.

COLOUR OF INTERNAL DECORATIONS WITH P.S.A.L.I.

Much can be done to avoid drawing the attention to the different colour rendering given by the natural daylight and by the artificial supplement, if care is taken in the choice of surface colours for walls, furniture, etc. in the room. The choice of such colours cannot very easily be specified because personal taste as well as functional requirement is to be considered. The following comments may be useful:

(1) Colours used on surfaces which have to extend from the area of natural daylight to the area receiving dominantly supplementary lighting (for example, the ceiling, cross-walls, or floor) should be chosen from colours which show the least change in the rendering given by natural daylight and by the chosen supplement.
(2) Areas of distinctive colouring should preferably not be extended, but should be confined either to the part of the room lit chiefly by natural daylight, or to the part of the room lit chiefly by the artificial supplement.
(3) Surfaces which receive the greater part of their illumination from the supplementary sources should be composed of colours which are not distorted unfavourably by the colour of the light source.
(4) The colour of the back wall in a single-side-lit room is a difficult case. Since the artificial supplement will be located near to this wall, it will receive a large component of artificial light but this will reach it as glancing incidence if the supplement is located in the ceiling. On the other hand, it will also receive a considerable component of daylight at near-normal incidence. A textured surface would immediately show up any differences in the colour rendering of the natural daylight and the artificial supplement. The only safe recommendation is to advise the use

of a neutral colour, equally rendered by both natural and artificial light, for use on the back wall.

APPEARANCE OF COLOURS UNDER DIFFERENT FORMS OF
FLUORESCENT LIGHTING

The colours of the British Standard B.S. 2660 have been examined experimentally under fluorescent light sources of different types. The changes which occur have been tabulated. (See Factory Building Study No. 8, *Colouring in Factories*, H.M.S.O., London 1961, Table 1.) The Table below is a digest of this information, relevant to the two types of fluorescent light source recommended at present for use in integrated installations.

TABLE
*Change of appearance of British Standard Colours under
'colour matching' and 'improved daylight'
types of fluorescent lighting*

(1) B.S. 2660. Colour change not generally acceptable.

Colour Matching	Improved Daylight
0–005, 006, 007	0–001 to 0–007
1–015, 016, 017	1–015 to 1–025
1–022, 023	8–090 to 8–092
8–090, 091, 092	

(2) Other colours are either not changed, or the change is hardly perceptible except under critical conditions.

The colours in (1) above can all be used, but should be confined to areas which do not extend from the daylit to the artificially lit parts of the room. The colours in (2) can be used without restriction except under critical conditions; special tests are then necessary to determine requirements.

The Table can only give a very rough guide, however, because the experimental work was done some years ago, and other forms of fluorescent lamp are continually being introduced with different colour-rendering properties. The only satisfactory recommendation is that specialist advice should be sought, and carefully controlled trials should be made, preferably in a scale model or in a full size room, of the combination of lamp and surface colour under consideration.

CHOICE OF SYSTEM FOR SUPPLEMENTARY LIGHTING

It will almost always be necessary that the supplementary lighting system should serve after daylight has faded, as part of the normal artificial lighting system. There is no reason why this should not be done, provided that the process of integration in the design is carried right through to the design of the artificial lighting as an entity.

The permanent supplement should not draw attention to itself during the day. The reason for this is simply that if the inhabitants of the building are aware of the artificial light sources during the daytime, it will be impossible for the supplement to give any impression of being part of the daylight system. On the other hand, if the light sources are suitably recessed, either in the ceiling or in a suspended laylight of suitable design, the daylight can still remain the dominant lighting feature of the room and can determine the character of the environment.

There is no one single system of supplementary lighting which has advantages over another. The inset laylight has been used and is a satisfactory solution where the room will be used to the greatest extent during the day and where the night time lighting is of secondary importance, and can be supplied by conventional fittings uniformly distributed about the room. On the other hand, trends in the design of supplementary lighting are towards a system which consists of inset lighting distributed uniformly over the whole ceiling, selected units only being used during the daytime to act as the permanent supplement. After daylight has faded, the whole installation is put into operation.

A third system consists of three parts of which one part is maintained permanently in operation. During the daytime units A and B provide a high level of supplementary lighting to the level required by the average daylight. At night time, system A is switched off and system C brought into operation, B and C

II

together providing a uniform level of artificial lighting over the whole room after dark. It is, of course, essential to draw blinds, of high reflection factor, over the windows at night both to avoid loss of light through the windows, to enhance the reflected component of the artificial lighting, and to give a visually pleasing environment.

It must be accepted, however, that any of these systems which incorporate switching arrangements to change the lighting at dusk will not be operated in the way intended by the designer unless the switches are controlled by the building management and not by the workers themselves. Left to themselves, workers will probably leave all the lights on all the time, either from laziness or inattention, or they may switch off some lights to satisfy some local glare problem or some individual preference.

P.S.A.L.I. AND 'OFFICE LANDSCAPING'

It may be useful to mention briefly at this point the technique, known as 'bürolandschaft' developed in Western Germany, which relies upon total artificial lighting over the whole area of a large office, but which includes some windows in the design primarily for the benefit of the view outside. The principles of P.S.A.L.I. are equally valid here since the sky seen through the windows will influence the adaptation. If the levels of artificial light are too low, the windows will cause glare; if they are too high, the outdoor landscape will appear dull, and on some days a weird and unpleasant effect will result. Levels of the order of 500–700 lux will generally be found best for office-landscaping installations with windows.

CRITICAL MINIMUM WINDOW SIZE

While it is a valuable feature of P.S.A.L.I. that it offers far greater freedom in the design of the windows than does total daylighting, there must clearly be some limit to the reduction in window size if the essential virtues of daylight are to be retained in a supplemented installation. Some experiments were undertaken to study this problem, and it was found that people judged the situation by more than one criterion. When they had freedom to vary the size and position of a window in a supplemented installation, they found that there was a certain critical minimum window size, but that their judgement depended upon whether they were assessing light quantity, or whether they were judging the view outside. In the latter case, their judgement depended greatly upon the nature of the view outside, how much of interest there was to see, and how much blank sky was in the view. It proved difficult to generalise from the experimental results.

Nevertheless there was considerable agreement about certain features of the experiment. When there is an interesting view outside, the critical minimum window size for a side window in a supplemented installation is of the order of one-sixteenth of the floor area. A window of less than this area has little more significance than a hole in the wall.

Since these experiments were done, further experience suggests that the result is a fair indication, and certainly a useful guide in design. It was an interesting feature of the study that the judgements were affected by the nature of the view, and this led to the suggestion that the shape, and position, as well as the size, might be critical in a different way, depending upon whether the view outside was of predominantly vertical lines (in a skyscraper town) or horizontal lines (rows of low buildings, or lines of gentle rolling hills). The answers to these and other questions awaits further research.

THE ROLE OF THE DAYLIGHT AND THE SUPPLEMENT
IN P.S.A.L.I.

Where the daylight and the artificial component in a P.S.A.L.I. system come from integrated rooflighting, both components play an equal and identical part. In a supplemented sidelighting system, however, the daylight component from the side windows plays a part which is different from the supplement, the difference depending very much upon the design and placing of the supplement. If the supplement is placed above the window in the window wall, as often in Scandinavia, or in the ceiling over the window, the supplement does no more than reinforce the daylighting in both quantity and in its modelling effect.

If, however, the supplement is placed in an entirely different part of the room, either distributed in the ceiling, or placed as a laylight over the parts of the room remote from the window, the role played by the two components of the integrated system are subtle and interesting. The side daylight will have a strong near-horizontal component which will light up vertical surfaces, but which will have only a small effect upon horizontal surfaces. A conventional measurement of illumination on a horizontal reference plane will show a low level of illumination, but the visual effect will be much greater because the eye notes and judges the lighting of the vertical surfaces—the walls, furniture, and human faces, and so this daylight component will have a visual effect far beyond that indicated by the horizontal illumination. Conversely, the supplement in the ceiling will appear to be much the more significant if reliance is placed solely upon the reading of a horizontal illumination meter, but the effect of the supplement will be chiefly on horizontal surfaces. Consequently even in areas of the room remote from the window, the daylight component will be playing a major or even a dominant role. This can easily be demonstrated in a P.S.A.L.I. system by blacking out the daylight and switching off the supplement in turn. In a well-designed system the former will have at least as great an effect upon the character of the room as the latter, even though the effect upon the working illumination may be greater at the back of the room when the supplement is switched off than when the daylight is removed.

The daylight component not only affects the judgement of quantity. Upon the modelling of solid objects, and especially of human faces, effects which are wanted and which are regrettably absent from conventional artificial lighting coming from the ceiling. A skilled designer of P.S.A.L.I. can make play with the combination of side and top light to achieve effects of a particularly pleasing kind.

Attempts have been made to quantify these modelling effects, and to express what is called the 'flow of light' in a supplemented system in terms of vector geometry. It is a simple matter to determine the vector resultant of two or more parcels of light flux, each of different amount and coming from different directions, but it by no means follows that this mathematical exercise relates to the behaviour of the eye faced with such a situation. The eye does not, in fact, see light 'flowing' in terms of such a vector concept. Just as the ear can sort out sounds of different loudness and direction for what they are in a way which is very difficult or impossible for a microphone array to reproduce, so the eye can sort out the light reaching an object into its components in a way which would be difficult to reproduce in a physical instrument. Reference has been made earlier to the effects of 'brightness constancy' (Chapter 3) and to the way in which the eye behaves when it can, and when it cannot, see the sources of light which are illuminating the objects of regard. In a P.S.A.L.I. system the eye makes judgements which are greatly influenced by what it can see of the whole system.

CALCULATION AND DESIGN OF SUPPLEMENTARY
LIGHTING

The detailed design of a supplementary lighting system is beyond the scope of a brief treatment of the subject. Basic design methods are given elsewhere (*Architectural Physics—Lighting:* Hopkinson, H.M.S.O., London 1963; and *Daylighting:* Hopkinson, Petherbridge and Longmore; Heinemann, 1966) and it is possible here only to outline the procedure.

The basic level of the supplementary lighting is first determined by analysis of the daylighting distribution (which may be predetermined for other reasons) or it may be sufficient to assume that the supplement can be designed to a level of the order of 500 lux as a first approximation. It should be borne in mind that unless some form of smooth variable control of the supplement can be installed to operate in sympathy with the changes in daylight, any fixed level of supplementary lighting is a compromise between the requirements strictly necessary for a wide range of outdoor sky conditions. It is therefore a satisfactory practical compromise to take the level of the supplement to be of the order of 500 lux except in special cases.

The selection of the supplementary system should next be made. This may for example be a laylight or a distributed system.

The distributed system can be calculated exactly as for an artificial lighting system which uses the same type of lighting unit, the coefficient of utilisation being determined in the same way. The calculation using the 'lumen method' is given in Chapter 5.

If a laylight, either suspended or inset in the ceiling, is used the calculation is more complicated, and reference should be made to Chapter 19, *Daylighting*: Hopkinson, Petherbridge and Longmore: Heinemann 1966. The procedure requires first that the size, proportions, and positioning of the laylight should be determined on architectural grounds. The larger the laylight, the less will be the likelihood of glare resulting, since the required amount of light can be distributed over the larger area and so give a lower brightness. The minimum area of laylight can be determined from an empirical formula:

Minimum area of laylight = area of working plane to be illuminated × required level of supplementary illumination ÷ 100.

The area of laylight obtained in this way is the smallest area which would not give rise to glare. An alternative limitation is to specify that the average luminance of the laylight must not exceed 1000 asb (100 ft-L approx.).

Once the area of the laylight is obtained, the rest of the calculation proceeds. A simple formula is given in Chapter 5 which can be used to determine very approximately the illumination given by a large source such as a laylight. Otherwise, the full design procedure (ibid, Chapter 19, Section 19.6.2) should be followed.

GLARE IN SUPPLEMENTED INSTALLATIONS

The degree of glare likely to be experienced in a combined day-light and artificial lighting system cannot be readily computed in terms of a Glare Index. The Glare Index for the artificial supplementary system can, however, be determined by the I.E.S. Glare Index method. If the Index so obtained does not exceed 13 or if the maximum luminance of a laylight does not exceed 1000 asb (100 ft-L approx.) it can then be assumed that any glare in the integrated installation will come from the sky only and not from the artificial supplement.

The sky brightness, unfortunately, is not under the control of the designer. The brightest summer sky can have a luminance of the order of 10 times that on a dull overcast day. Hence the Glare Index from a given window can vary widely from occasion to occasion. There is no remedy but to provide either a fixed anti-glare system which screens off the greater part of the sky from observers' most sensitive positions in the room, or to provide adjustable blinds or screens to be used on excessively bright days. While there is no basic difficulty in computing a Glare Index for a window or system of windows, relative to a given viewing position in a room, the exercise is of more theoretical than practical value. It is better to design the fenestration, integrated with the artificial supplement, to give a satisfactory illumination and brightness distribution over the room, providing such permanent or adjustable anti-glare devices as are desirable for the particular environment (for example a hospital ward) and to design this system with reference to the average sky conditions. For excessively bright skies, additional blinds, curtains or screens should be used. If the design is based upon the brightest sky likely to be experienced, it will probably be unduly restrictive in relation to a view outside on average and dull days. Equally, if the design was based on conditions on dull days, the blinds would have to be pulled on most days of the spring, summer, and autumn.

7

The Lighting of Schools

School design periodically undergoes radical changes, partly as a result of new ideas on education and partly from developments in school building techniques. Such a revolution in design has been a feature of the last twenty years. The need to build a large number of new schools to house a growing population and to replace the old church schools and Board schools has been the stimulus to new thought on school building. The educational system of this country was defined in the 1944 Education Act, the chief point of departure of which was the creation of a system of secondary education for all children of the ages of eleven upwards which demanded the erection of a new set of school buildings designed initially around the concept of seconday modern education and later around the concept of general comprehensive secondary education.

LIGHTING AND THE ONSET OF VISUAL DEFECTS

One of the chief causes of complaint against old school buildings was their inadequate lighting whether by day or by night. During the inter-war years, growing concern had been expressed by educational and by medical authorities at the incidence of defective vision in school children, particularly of older school children. The onset of defective sight appeared to correlate so clearly with the increase in the intensity of school studies that most informed opinion believed that there was a causal relation between the two. Bad lighting was felt to be chiefly responsible for this unfortunate state of affairs.

168

Great efforts were therefore made to investigate the relation between lighting and vision of schoolchildren and to see in what way lighting could be improved. These studies were coordinated by a Joint Committee on Lighting and Vision set up by the Building Research Board of the Department of Scientific and Industrial Research, and the Medical Research Council. This Committee initiated, fostered, and coordinated research on the vision of schoolchildren in relation to lighting conditions.

The most significant research project was that of Sorsby at the Royal College of Surgeons. Sorsby demonstrated by a series of experiments that the onset of myopia, the chief progressive visual defect in schoolchildren, and the one for which bad lighting was considered to be responsible, was basically a hereditary condition, the development rate of which might be influenced in minor degree by nutrition or other environmental factors, but which could not in any major degree be attributed to lighting conditions. This indicated a state of affairs quite different from that which had been feared. Subsequently Sorsby's conclusions, at the time revolutionary, have been confirmed in the sense that there is no evidence that children working in new school building with excellent lighting are in any way more free from these visual defects than are children working in older and worse lit buildings.

The researches at the Building Research Station demonstrated, however, that better lighting unquestionably led to better visual performance and particularly to more accurate and easier reading. The result of these two researches was clear. First, school-children and their teachers would undoubtedly benefit by better lighting, because they would be able to see more easily and would perform more accurately, and would consequently probably be less tired at the end of the day, but equally it was clear that better lighting was not likely to halt the progress of those visual defects which were associated with growth and heredity. Once these facts were established, there was a sounder basis for the provision of good lighting while at the same time no false expectation was held out that children's eyesight would improve as a result of better lighting.

Parallel with these laboratory researches, development work on the design of schools had demonstrated clearly that better lighting led to a feeling of freshness and buoyancy and that pupils and teacher alike appreciate the creation of a pleasant and stimulating visual environment of appropriate character. Both daylighting and artificial lighting were equal partners in the creation of such a pleasing environment and those responsible for school development insisted that artificial and natural lighting should be integrated with the design of the building as a whole. This was a change from the previous pattern in which the architect had considered himself to be responsible for the daylighting but had handed over the artificial lighting to be added as an afterthought by an electrician with little or no appreciation of the basic problem.

CRITERIA FOR GOOD VISION IN SCHOOLS

The visual acuity of children (that is, the smallest detail that can be seen expressed in terms of the angular subtense at the eye) is not significantly different from that of adults, but being less experienced, children are not able to interpret visual information with such certainty. An adult can interpret a slightly blurred image from experience, for example, he can read words accurately even though he may not be able to see the individual letters because long experience has taught him the shape of words and the sequence in which they appear in a normal sentence. A child who is learning to read letter by letter or by the word-picture method cannot interpret in the same way. Consequently the vision of school children, though objectively at least as good as that of adults, is subjectively considerably inferior and so it is necessary to ensure that their visual tasks are made very much easier.

The majority of schoolchildren have normal vision but their visual performance is related to the amount of light in which they work. The relation between visual performance and level of illumination has been studied in great detail and it has been established that if the illumination on normal school visual tasks is less than 10 lm/ft² (about 100 lux) visual performance falls off rapidly and this, therefore, is the minimum illumination which should be tolerated on visual tasks in a schoolroom, whether by night or by day. As a result of the work at the Building Research Station, the education authorities laid down in school building regulations that 10 lm/ft² should be the minimum level of illumination in all teaching areas.

Children benefit from levels of lighting in excess of 10 lm/ft² provided that such increases are not accompanied by unwanted glare either through excessive view, bright sky, or of direct view of bare lamps. The Illuminating Engineering Society recommends an average level of 300 lux (30 lm/ft²). Levels higher than this are recommended in the United States of America.

Children with subnormal vision benefit rather more from increased lighting levels than do children with normal vision. It is, however, rarely possible to overcome any visual defect merely by the provision of more light. A combination of good lighting and proper correction of the visual defect is essential, where the defect is a refractive error which can be corrected by spectacles. While such children benefit slightly by the provision of higher levels of lighting, they benefit far more by being allowed to sit near to the chalkboard. Children with defects so serious that they require special assistance in their education will be catered for either in special classrooms in a normal school or in special schools for the partially sighted. Schools for the partially sighted will be discussed later.

It has been shown earlier (Chapter 3) that visual performance depends upon the size of critical detail and critical contrast. One of the more difficult visual tasks which children have to undertake in school is reading the chalkboard. The factors here are the distance of the child from the chalkboard, the size of the letters, and the contrast of the chalk lettering with the background of the chalkboard. It was once believed that a deep black chalkboard with sharp white letters was essential for good visibility. This is true, but not to the extent which was once believed.

The most comfortable condition is one in which the letters are seen against a medium grey rather than a dark black, because white letters appear to 'dazzle' less with the slightly lighter background. The use of coloured boards with white chalk lettering is not only fully acceptable so far as clarity of vision is concerned, but may also be positively indicated to enable the chalkboard to be a natural focus of attention.

The researches at the Building Research Station referred to earlier demonstrated that the real criterion in the determination of lighting in schools is not only the ability of the child to see individual detail in terms of size and contrast, but also his ability to read quickly and accurately in relation to his age and educational attainments. The Building Research Station work distinguished between basic visual acuity, speed and accuracy of reading, and ease of reading. It was demonstrated that in a group of people with 'normal' vision, some would be able to read more quickly and accurately than others, and that others would be able to read easily where some might have difficulty. Clearly what was involved here was not only the basic optical equipment of the eye, but the interpretative faculty of the individual (see Chapter 3). It can therefore readily be understood that older children who are learning rapidly and who already have a considerable store of knowledge will be able, by making a visual effort, to interpret material seen indistinctly on the chalkboard, whereas younger children with the same degree of visual acuity, seeing something quite unfamiliar for the first time, will let it slip by if they cannot see it clearly. The researches showed that these factors can be taken into account better by increasing the apparent size of detail rather than by lighting it to a higher level. It was shown that under common classroom conditions, it can often happen that moving a child three or four feet in towards the chalkboard can produce the same improvement in visual performance that would be given by raising the lighting level by a factor as great as thirty times. This makes it clear that there must be an intelligent appreciation by the teacher of the need to write large clear letters, and that most certainly it is not possible to correct, by the provision of high levels of lighting, troubles which arise through overcrowded classrooms which relegate children to positions too far from the chalkboard (Fig. 6).

A different situation arises in relation to near vision tasks. The distance between the eyes and the work is under the control of the child himself and it will inevitably happen that children with relatively poor vision will, in order to see better, tend to reduce the distance between their eyes and the reading material. Good lighting can certainly help to correct this, but because so much greater improvement results from quite a small reduction in distance, than results from increasing the level of lighting, the tendency will always be for the children to look too closely at their work.

LIGHTING THE WHOLE ENVIRONMENT

The implementation of the 1944 Education Act has had a major effect on school building design. The tendency has been away from rigid teaching methods and therefore from rectangular plans for classrooms, towards freer space for more flexible teaching methods. The class, instead of living its whole school life in rows of desks, spends much of its time in small groups each engaged in different activities. Everything is done to adjust the scale of the classroom to the size of the child. Ceilings have been lowered, furniture has been related to the size of the child, and everything done to give a more intimate character to the school. These trends in planning have been most marked in schools for young children, but equally in the provision of new secondary schools, a great deal of new thinking has been necessary. Less teaching is done at the chalkboard and much more by small groups working at times by themselves and at times with a teacher. Classrooms tend therefore to be planned as multi-purpose spaces with the consequent need for good lighting all over the teaching space. Lighting and planning have to go hand in hand from the start.

Light and colour as well as architectural form create the

character of the visual environment. It was once thought that a school should have an austere and institutional character consistent with the need to impose and maintain discipline. High windows which admitted light but which did not permit a view outside, dark walls up to finger height to hide the dirt, and dark floors to mask the dirt from muddy boots were characteristic of this approach to the problem. All this has changed. The school designer has attempted to make a positive contribution by creating a pleasant and intimate environment in which the child feels at ease, and which beckons rather than repels him.

Dull uniformity of lighting is not stimulating and so must be avoided in a classroom. All teaching space must have adequate lighting, and there should not be excessive contrast between the lighting of one place and another. Dull uniformity can be avoided by the judicious choice of reflectance and colour to ensure that some areas have a different character from others. There is advantage to be gained also in preferentially lighting certain focal areas of the classroom, including, of course, the chalkboard, or in having additional lighting over particular areas of special activity such as a book corner or a display wall.

Harsh changes in brightness and illumination must be avoided, however, because eyes do not adapt instantly from one level of brightness to another. If a child has to look up to a very brightly lit chalkboard and then back again to a poorly lit book, re-adaptation may be slow and visual performance less than would be indicated by the available amount of light on the book. Lighting and decoration should therefore be so arranged that the eye never has to make big adaptive changes. Contrasts in brightness (luminance) greater than 10 to 1 should be avoided in juxtaposed areas, and as far as possible the relation of the brightness of the work to its immediate background should not be greater than 3 to 1. In practice, this means, for example, the provision of desks and work tops with a reflectance of the order of 25–30%, that is, about one-third of that of white paper.

ATTENTION, DISTRACTION AND GLARE

The distribution of light in a school has a profound effect on the way in which the visual attention can be held or on whether it is likely to be distracted. The eye has a natural tendency to look at bright objects. The attention is also taken by contrasty and colourful objects, and by things which move in the field of view. Children's eyes will irrevocably be attracted to bright areas, whether these are the things at which they should be looking or not. The older type of school classroom with its high windows deliberately placed to avoid giving a view outside did not succeed in their purpose because they were a sure distraction because of their high brightness. When a child in a gloomy room sits looking apparently longingly through the window at the sky, it is responding naturally to a phototropic impulse. The only way to come to terms with this phototropic property of the eye is so to light the room that the things to which the eye naturally gravitates are also those on which the teacher wants to hold the children's attention. Clearly this precept cannot be carried to its limit because otherwise windows would be obscured throughout the day by blinds, and the chalkboard would be floodlit. Nevertheless by taking this phototropic property of vision into account, much can be done to ensure that attention is held without strain and that distractions are kept to a minimum. Preferential lighting on the chalkboard, and the use of a coloured chalkboard which contrasts favourably with its less distracting surroundings, are obvious ways in which the attention can be held. There are obvious faults to be avoided. Teachers often place their desk at an angle to the class with their backs towards a window. This is often the most comfortable visual position for the teacher, but inevitably it means that the distraction to the children of the bright window is overpowering and any attempt to resist this distraction must result in some form of visual or general fatigue.

Areas of excessive brightness in the visual field of the classroom can also cause glare. The teacher sitting backing a window will not be seen clearly by the children because the bright sky will cause a certain degree of 'dazzle' or disability glare. Large areas of bright sky also cause discomfort, as does the direct view of bare lamps in lighting fittings. The chief forms of control which the designer has for the elimination or the reduction of glare from natural lighting are adjustable blinds and curtains. The adjustable venetian blind, provided it is well designed and is properly used, gives almost complete control of sky glare and sun glare. By intelligent manipulation of the angle of the slats, daylight can be allowed to penetrate towards the back of the classroom, particularly after reflection from the ceiling, while still eliminating direct view of the sky especially to children near the windows who are the greatest sufferers. Unfortunately blinds are not always adjusted properly, and this has led to consideration of fixed louvres, either horizontal or vertical, for the control of sun and sky glare. This is also unsatisfactory because on dull days they cut out wanted light.

Other forms of light control such as the use of translucent pleated paper blinds and woven translucent plastic roller blinds, all have their advantages. Woven plastic blinds can be made in such a way that when drawn, there is still a view outside. Such blinds are not a sufficiently satisfactory screen to direct sunlight, but reduce sky glare. Pleated paper blinds have the advantage that they add substantially to the thermal insulation of the window when the blinds are drawn. At night any light-coloured blind can offer a good reflecting surface to assist the artificial lighting.

There is much to be said for the use of white washable draw-curtains. A translucent white curtain which makes only a little reduction in the internal daylight in a room may result in a major reduction in sky glare. Their initial cost is low, and their only disadvantage is a relatively short life and the risk of lack of proper laundering and maintenance.

Sky glare can to some extent be reduced by the introduction of reveals to the window which grade the contrast between the

ART GALLERIES

In most art galleries a chief aim in the design of the lighting is that the exhibits should appear the same—as near as possible—by daylight, by artificial light, or by a combination of the two. To achieve this needs careful control of the distribution and colour rendering of both natural and artificial light sources.

Plate 39 (*a and b*).

Section 1. Egg crate velarium
2. Tungsten spot lights
3. Fluorescent lamps in parabolic reflectors
4. Tapered wooden fins
5. Cornice re-shaped to avoid direct daylight
6. Suspended metal vanes to reduce daylight on floor
7. Tungsten fittings
8. Roof glazing with electrically-controlled cloth roller blinds

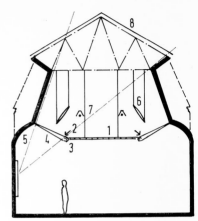

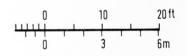

(*b, below*). The City Art Gallery in Birmingham was remodelled in 1957. The upper part of the building was re-shaped to concentrate the maximum illumination on the picture walls, and a velarium was introduced to cut down the view of the sky and to reduce brightnesses in the centre of the gallery so that reflections in picture glass were minimised. (Architects: A. G. Sheppard Fidler in conjunction with the Building Research Station. Photograph by courtesy of the City of Birmingham.)

Plate 40 (a, b and c). The 19th-century building of the Whitworth Gallery in Manchester University has been re-modelled between 1963 and 1967, with a new floor inserted to give additional display space. The daylighting and artificial lighting has been designed to give a level of illumination of not more than 12–15 lm/ft^2 (130–160 lux) in order to conserve the collection of water colours and textiles. The artificial lighting is from a mixture of approximately 2/3 colour-matching fluorescent lamps and 1/3 tungsten lamps, and is designed to be used in combination with the daylighting when required. The proportion of fluorescent to tungsten lighting can be adjusted. (*a, opposite*). New mezzanine level in Margaret Pilkington Room. (*b, above*). View from Calouste Gulbenkian Room. (*c, below*). Section: 1. Roof glazing. 2. New windows in some positions with adjustable cloth/louvre blinds. 3. New flush ceiling of Sitka spruce with louvred apertures for daylight and fluorescent lighting, and with recessed tungsten fittings. 4. New mezzanine floor and staircase. 5. Movable display screens. 6. Close-louvred ceiling of Sitka spruce with cross louvres above. (Architect: John Bickerdike. Photographs by Elsam, Mann & Cooper.)

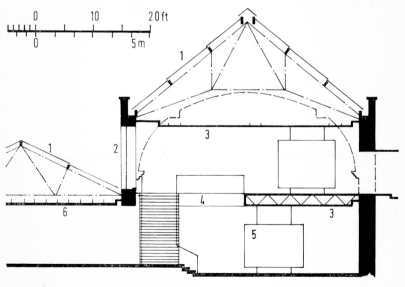

Plate 41. The gallery in the Commonwealth Institute in London is a room 95 ft × 45 ft × 13 ft high (29 × 14 × 4 metres high). The roof structure forms deep louvres to the roof lights which give an even level of illumination over the whole gallery. A low level window gives a view out at one end. Colour-matching fluorescent lamps are mounted in the roof lights. (Architects: Robert Matthew, Johnson-Marshall and Partners. Photograph by Henk Snoek.)

Plate 42 (a and b, opposite). In Room VII at the National Gallery the needs for conservation called for illumination not exceeding 15–20 lm/ft² (130–160 lux) on the pictures. Daylight is controlled by motor-driven venetian blinds monitored by photo-electric cells (see *a*). Failing daylight is supplemented by fluorescent lamps above prismatic panels which direct the light on to the walls, with a dimmer circuit again controlled by photo-electric cells. Ultra-violet light is excluded by the special acrylic sheet ceiling and by cellulose acetate lamp sheaths. (Photographs Crown Copyright Reserved.)

Plate 43. Kasmin Gallery, London. In this con
cornice has been inserted at an angle which ens▸
surface is the picture wall. Deep plaster baffles redu
light, and adjustable fabric louvres control sunlig
light the gallery after dark, with additional spotligl
to the aluminium booms. The walls are white texture
covering is light grey ribbed rubber. (Architects:
Koralek. Photograph by John Donat.)

bright sky and the interior of the room. Such contrast grading is always to be advocated, even though on very bright days it will always be necessary to draw the blinds or curtains.

In the same way, the window wall should always be painted a very light colour in order to reduce the danger of excessive contrast with the sky. The use of dark colours near the window must be avoided at all costs, whatever other argument there may be in favour of the use of such colour. This is an occasion, and there are many, to reiterate that the reflectance design, that is, the interior decoration, of a room is as much a part of the total environmental design for visual comfort as is the design of the windows or of the artificial lighting fittings. Any redecoration scheme must always take this into account.

THE IMPLEMENTATION OF LIGHTING STANDARDS—NATURAL AND ARTIFICIAL LIGHTING

Standards of lighting have been laid down by the Department of Education and Science which are directed towards ensuring that adequate lighting will be provided in school buildings to enable all visual tasks to be accomplished easily and well and to give visual conditions which are comfortable, particularly avoiding unnecessary glare discomfort. These standards have to be implemented in practice in terms of both natural and artificial lighting. The problems are essentially different in the two cases. Natural lighting comes from a large area, the sky, and varies in amount throughout the day and the seasons of the year. In addition, sunlight comes from a small part of the sky, changing in direction but varying very little in intensity. Artificial lighting, on the other hand, comes in a fixed amount which is determined at will by the designer.

The provision of both natural and artificial lighting is determined by economic considerations. With artificial lighting the capital cost of the lighting fittings and of the installation, the running costs in terms of electric current and of maintenance, cleaning, etc. limit the amount of artificial illumination which can be provided. In the case of natural lighting, the balance has to be struck between a number of factors, of which the loss of internal heat during the winter through glass, and the greater cost of window area as compared with wall area, together with the problem of sound insulation from unwanted noise in urban areas, are probably the most important. As a result of a much better understanding of the need for good lighting, it is very rare nowadays that economic considerations result in nothing more than the bare minimum standards of lighting. This is because designers above all recognise that good lighting is one of the most important aspects of environmental design, and they are prepared to make economic sacrifices to achieve it.

The Department of Education and Science lays down that there shall be a minimum daylight factor of 2% in all teaching areas. This regulation, as now framed, derives directly from the research work on lighting and vision of schoolchildren conducted at the Building Research Station under the aegis of the Joint Committee on Lighting and Vision of the Medical Research Council and the Building Research Board. Prior to this research having been completed, however, previous regulations of the then Ministry of Education were based on different concepts, which governed much school building in the early post-war years and whose influence is still felt where modern ideas have not yet penetrated.

At the time that the earlier regulations were enforced, no simple means existed for calculating the reflected light in a room. Consequently the term 'daylight factor' was understood then in terms of only the direct light from the sky. This led inevitably to gross overglazing of school classrooms. Architects found that the only way in which they could implement the regulations as they were then interpreted was by the use of windows in opposite walls, the usual arrangement being to employ a large main window associated with a high clerestory on the opposite wall. This in turn was allied with the dispersed or 'finger-plan' type of classroom arrangement with teaching areas on one side of a long corridor. This planning was unsatisfactory because it was un-economical in the use of site space, and almost unavoidably led to long depressing tunnels of corridors with classroom doors ranged along them, giving the school an overbearingly prison-like character. The architectural result of this approach to light-ing carried to excess of overglazing had not been anticipated by those responsible for the original regulations. Once the errors of overglazing and spider-like planning had been appreciated, architects sought other solutions which would enable the light-ing requirements to be met while at the same time leading to more satisfactory planning in terms not only of single-storey buildings but also of multi-storey buildings.

In parallel with this architectural reappraisal, the situation was being examined both from an education and from a lighting point of view. The educational reappraisal called clearly for more compact planning leading to a more intimate character of the school building as a whole as well as of the individual class-rooms. New ideas of group teaching were difficult to implement in a dispersed plan, and the overglazing of the rooms made it impossible to think in terms of local areas of special activity because the enormous windows dominated the whole character of the room.

From the lighting side, the researches at the Building Research Station demonstrated that the very high levels of lighting pro-vided by these overglazed schools were unnecessary and were definitely unsatisfactory because of excessive sky glare. The over-glazing was also leading to excessive intake of solar radiation causing heat exhaustion on sunny days, while in the winter there was loss of internal heat and cold draughts and low internal temperatures near to the windows.

The building regulations of the Department of Education and Science as they now stand, therefore, require a minimum day-light factor of 2% in a teaching area, the term 'daylight factor' having the meaning agreed internationally, that is, the ratio of the total internal daylight, direct skylight and all reflected light taken together, to the total external available daylight. Provided full advantage is taken of internally reflected light, glazing can be reduced considerably over that thought necessary when reflected light was ignored.

The area of glazing necessary to achieve the minimum daylight factor of 2% will depend upon the shape and position of the windows, whether or not the glazing is in more than one wall, and upon the lightness of the internal decorations which govern the internally reflected component of the total daylight. In turn the design and arrangement of the necessary glazed area in the form of windows must depend greatly on the way in which the teaching area will be used. In most primary schools and many secondary schools children will work often in groups facing in different directions and so the disposition of windows will not necessarily be related to one particular direction of viewing. This makes design more flexible and also more difficult. For example, in many countries on the Continent where formal teaching in rectangular classrooms is still common, windows are placed in classrooms to ensure that all daylight comes in from the left (the assumption is that all children will write with the right hand) and so unilateral lighting is the simple and invariable rule, even though the placing and design of the windows in the form of stepped sections may show considerable ingenuity.

The penetration of daylight into a room has been discussed in Chapter 4. Daylight on a school desk is thought of in terms of the two components, the sky component received through the window directly from the sky, and the internally reflected component which is the light reaching the point after having been reflected and interreflected from all the surfaces in the room. The internally reflected component can be regarded as a supplement of constant value throughout the whole room to be added to the sky component, which varies with the distance of the point from the windows. The internally reflected component ls determined by the average reflectance of the internal room surfaces, and by the area of glazing. Consequently all the room surfaces will have an influence on the internally reflected component, not forgetting the floor which receives, and therefore reflects, much of the light coming in directly from the sky.

The area of glazing necessary to achieve the required quantity of daylight in a classroom depends upon the type of design solution which is proposed. When side-lighting alone is employed it is impossible to achieve the statutory quantities of daylight if the glazing has a total effective area of less than one-eighth of the floor area. Even so, it is necessary that the glazing be placed in more than one wall. If the glazing is in one wall only, it is difficult without recourse to very high ceilings to achieve the necessary quantities of daylight throughout the classroom with glazed area of less than about one-fifth of the floor area. This means in effect that most of the window wall, less structural parts and glazing bars, must be glazed from the sill up to near ceiling height.

The introduction of top lighting, even a small amount, to supplement the light in the parts of the room remote from the side windows, can substantially add to the effectiveness of the glazing, but even so, it is difficult to achieve the necessary lighting levels with glazing of less than one-tenth of the floor area if a satisfactory view outside is also to be provided. It must be remembered, also, that this glazed area is the area of unobstructed glass, and not the area of the window opening in the wall.

On the other hand, there is no need to use glazed areas in excess of one-quarter of the floor area. The building regulations can be met, even by unilateral lighting with a ceiling height no greater than 9 ft., with a glazed area of no more than one-quarter of the floor area in a classroom to accommodate forty places, provided the ceiling is of a light colour of reflectance not less than 70%, the walls not less than 50%, and, most important, the reflectance of the floor not less than 40%. Furniture must be kept light, with desk tops of a reflectance of not less than 30%.

WINDOWS, GLARE AND ORIENTATION

It is always preferable, where possible, to place windows in more than one wall, and preferably in adjacent walls. There are advantages in this not only as regards daylight penetration. It is easier to ensure adequate daylight penetration if windows can be placed to give preferential lighting to places which require it, and particularly to enable local areas of special activity to receive preferential daylight. (It must be remembered that the regulations call for not less than 2% daylight factor over the 'teaching area'. An area near a door may not be used for teaching and can therefore receive less than the statutory minimum.)

Windows in adjacent walls alleviate glare because the window in the one wall lights up the other wall and so reduces the contrast between sky and interior. This is true whether the window is in an adjacent wall or in an opposite wall. The disadvantage of lighting from opposite walls is that it gives rather troublesome modelling, and due to the phototropic effect, the eyes are attracted first one way then the other in a way which can be fatiguing.

The orientation of a school building has little effect on the quantity of interior daylighting because in the climate of Great Britain and of temperate, maritime countries in general, the sky is often overcast and the light from different parts of the sky is only slightly influenced by the position of the sun on such overcast days. Southern skies are sometimes brighter than northern skies, but the differences are marginal and are always neglected in daylighting design in cloudy climates.

The main effect of orientation is in relation to the advantages or disadvantages to be gained by direct sun penetration. The penetration of direct sunlight in a room can create a feeling of well-being, provided that the sunlight is under control. There is much to be said for arranging the orientation of a building so that every room receives sunlight on some occasions. A room orientated directly to the north will never receive sunlight during school hours. This may not be a great hardship, but most of the teaching staff would agree that sunlight is an amenity which they like to have, provided that they can control it. In practice this means that classrooms should be so designed and orientated that windows can be placed in side walls facing east, through south, to west, so that at any time of the year the sun, if it is not obscured by cloud, can penetrate into the classroom for a while at least. No standards for sunlight penetration are laid down in the school building regulations. However, the standard accepted by many local authorities for dwellings, that sunlight should penetrate for not less than one hour during all the year except for the months of December and January, can usefully be taken as a standard for school building.

The same devices used to control sky glare—venetian blinds, adjustable curtains, etc.—can be used to eliminate unwanted sunlight. On the other hand, the build-up of excessive solar heat in a classroom cannot be effectively controlled in this way. It is always advisable to lower the blinds when the sun is shining too strongly directly on the children because this will reduce the direct solar radiation, but internal blinds have only a slight effect on temperature build up, which must be controlled by avoiding excessively large windows facing the sun, or by the use of exterior blinds or canopies to prevent the solar radiation from entering the room in the first place.

ARTIFICIAL LIGHTING IN SCHOOLS

The standards of illumination necessary for satisfactory vision in school teaching areas apply equally to the artificial lighting as to the daylighting. The school building regulations of the Department of Education and Science lay down that a minimum level of illumination of 10 lm/ft^2 over the teaching area should be provided from artificial sources. Although in many schools, especially primary schools, and particularly in southern England, artificial lighting will only rarely be used, nevertheless an installation must be provided which is satisfactory in every respect.

The statutory minimum of 10 lm/ft^2 is not a design standard.

It is what it states, a minimum below which the lighting level must never fall. The design standard should be much higher. The Illuminating Engineering Society, a professional body responsible for setting lighting standards in this country, recommends an average level over the classroom of not less than 300 lux (30 lm/ft² approx.).

The school building regulations also lay down that artificial lighting fittings should be so designed that they do not cause troublesome glare. At present the regulations are framed in such a way as to specify a maximum tolerable brightness (luminance) of the lighting fitting as seen from a normal viewing angle, that is, that this brightness should not exceed an average of 1000 ft-L (about 1000 asb). The purpose of this regulation is to reduce glare discomfort from the artificial lighting system to an acceptable level. The Illuminating Engineering Society recommendation is in terms of the Glare Index, stating that the Glare Index of the installation in the classroom should not exceed a value of 16. These two recommendations, the first mandatory from the relevant government department, the second in the form of a recommendation, are basically consistent, but the mandatory regulation can be administered by direct photometric measurement of the lighting fittings in a classroom (Fig. 16).

The performance of an artificial lighting installation in a school classroom will depend upon the reflectance characteristics of the walls, ceiling, and floor in the room which in turn, as we have seen, are an integral part of the daylighting design. The lighting fittings must be designed to distribute their light in such a way as to prevent excessive contrast between the fittings and their background. In effect, this means that the lighting fittings must emit light in all directions, and not only down on the desks. Light must be distributed around the ceiling and walls to ensure a satisfactory measure of brightness in the surroundings, since this surrounding brightness is necessary in order to buffer the glare from the lighting fittings themselves.

In the design of a lighting installation for teaching rooms in a school one of the main problems which the designer has to solve

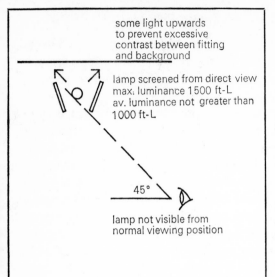

some light upwards
to prevent excessive
contrast between fitting
and background

lamp screened from direct view
max. luminance 1500 ft-L
av. luminance not greater than
1000 ft-L

45°

lamp not visible from
normal viewing position

Fig. 16. Light fittings for schools. The diagram illustrates the points required by the current (1959) D.E.S. Regulations to ensure comfortable viewing conditions. It relates to a fitting with a luminous area of not more than 100 sq. ins. (0·06 sq. m.). (Diagram by courtesy of H.M.S.O.)

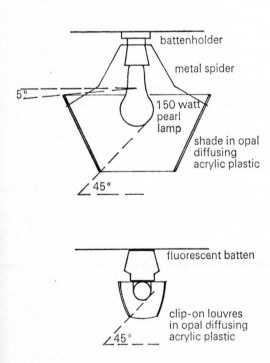

battenholder

metal spider

5°

150 watt
pearl
lamp

shade in opal
diffusing
acrylic plastic

45°

fluorescent batten

clip-on louvres
in opal diffusing
acrylic plastic

45°

is the choice between incandescent filament lighting and fluorescent tubular lighting as his source of illumination. This choice is partly one of economics, and partly one of preference (see Chapter 5).

The filament lamp is a cheap and simple light source, easy to maintain and to replace, familiar in its operation, and of a colour which we associate with warmth and domesticity. It has a disadvantage that its efficiency as regards conversion of electrical energy to light is low and therefore the running costs of a filament lighting installation are relatively high as compared with the much more efficient fluorescent lamp. The disadvantages of the fluorescent lamp were once that it was unfamiliar and its colour was unpleasant and institutional in character and totally unsuitable for the intimate and colourful character of the classroom as striven for by the designer. These objections are becoming less potent as fluorescent lighting becomes more familiar and major improvements are made in its colour characteristics.

The choice between fluorescent and filament lighting in school classrooms is therefore a choice between a light source with low initial cost, simplicity in operation and maintenance, and familiarity, but with high running cost and a source with a high initial cost of both lamp and associated ballast gear and fitting, but with low running cost.

If the artificial lighting will not be used outside normal school hours (apart from the extra hour for closing down, cleaning, etc., it may be difficult to justify the higher capital cost of fluorescent lighting. Once these conditions no longer apply, however, the economic advantages of fluorescent lighting become progressively greater the higher the level of working illumination and the longer the hours of use. In practice this means that filament lighting is indicated in primary schools which are not used in the evenings, but in secondary schools where the hours of work are longer and which are often used in the evening, the case is in favour of the installation of fluorescent lighting.

It is advisable, if fluorescent lighting is to be installed in teaching areas, that it should also be installed throughout the school.

It is inadvisable to create a situation where children will pass from a classroom with a high level of fluorescent lighting to a corridor or cloakroom with a low level of filament lighting.

Choice between filament and fluorescent lighting is also a question of design. Filament lighting lends itself to the design of small compact and tidy lighting fittings which are not obtrusive during the daytime. Unfortunately many fluorescent fittings are bulky and are more obtrusive by day, and also by night, than is desirable. Fluorescent lighting can of course be well designed and properly integrated with the structure of the building.

Fluorescent lighting has now been in use for over twenty years, and there is no evidence that it is in any way injurious to sight. Originally there were complaints and a great deal of prejudice against fluorescent lighting because badly designed and badly maintained fluorescent lighting caused complaints on the grounds of inadequacy, glare, and flicker. There are certainly some people who are abnormally sensitive to flicker, possibly 1% of the population, and such people may experience headache when working under fluorescent lighting. They will also be unable to watch television (a useful test for malingering). So far as is known, even these sensitive people will suffer no more than temporary discomfort, and any child or member of staff who appears to suffer real distress when working under fluorescent lighting should, of course, submit to a thorough medical examination because if there is a fault, it will be with the person and not with the lighting. In any case, however, these special problems of fluorescent lighting are likely to disappear as lamps and their associated electrical control gear are improved to give a light source which does not fluctuate appreciably with the alternations of the electrical supply.

Lighting fittings for use in schools have to meet the statutory building regulations for maximum average luminance and screening of the bare lamp, and must also be easy to maintain in a clean condition (Plate 44 a and b). Simple, basic designs for both tungsten filament and fluorescent lamps have been prepared by the Department of Education and Science in collaboration with the Building Research Station, and variants and develop-ments of these and other designs are available from manufactur-ers of lighting equipment. These basic designs were intended to provide the recommended levels of illumination within the limits of freedom from glare, for installation in classrooms of not more than forty places, and on the assumption that the reflectance characteristics of the room surfaces would be those necessary for adequate daylight design as discussed earlier. In large rooms (1000 sq. ft. and upwards), and in special areas such as teaching laboratories, and indeed in any room where an average level of illumination above the I.E.S. recommended level of 300 lux is to be provided, other types of lighting fitting with a somewhat greater proportion of downward light in relation to side light must be chosen if the I.E.S. limiting value of Glare Index 16 is not to be exceeded. Linear fluorescent fittings should be mounted with the long axis of the fitting in the direction of viewing and not across the direction of viewing because glare discomfort with such linear fittings is greater when they are viewed side-ways than when they are viewed lengthways.

When the choice of light source and of lighting fitting has been made, the number of lamps and fittings to be provided can be calculated by any of the well-known methods (see Chapter 5).

Lighting calculations will show that, to a sufficient order of approximation, between three and four watts of filament lighting per square foot of teaching area will be required, and about one and a half watts of fluorescent lighting per square foot. This assumes reflectance properties of the wall, ceiling, and floor sur-faces to give an average reflectance of the order of 40%, and it is also assumed that lighting fittings of the type which meet the building regulations, similar to the Department's own design of lighting fittings, will be employed. The calculation also assumes that light coloured blinds will be dropped over the windows at night, to aid reflected light.

Thus for a classroom of area 600 sq. ft. about nine filament fittings each of 200 watts output or 12 fittings of 150 watts out-put will be satisfactory.

Fluorescent lighting would normally be designed to a higher level of working illumination because it would usually be installed in secondary schools and other areas where higher working illumination is indicated by the more difficult visual work. A secondary school classroom of 900 sq. ft. could be adequately lit by twelve fittings each containing one 80 watt fluorescent tube or twin 40 watt fluorescent tubes, or alternatively six fittings with twin 80 watt tubes, arranged so that the fittings are seen end on in the main line of sight such as when viewing the chalkboard. The same number of lamps would be necessary in a system where the lighting is integrated into the structure.

LOCAL LIGHTING—THE CHALKBOARD AND OTHER AREAS

Preferential lighting is required on all areas which demand special visual attention. In a conventional classroom the chalkboard is such a special case. It is desirable that the level of illumination on the chalkboard should be greater than that on the surrounding wall in order to command attention and also to give a greater ease of seeing. Special directional lighting fittings should be provided which give the necessary level of illumination without spilling light unnecessarily on to the surrounding wall. This spill creates an area of distractingly high brightness around the chalkboard and therefore defeats the purpose of the selective local lighting. The level of this supplementary illumination on the chalkboard should be of the order of 600–1000 lux (60–100 lm/ft²) and certainly not less than 300 lux (30 lm/ft²). It can usefully be used during the daytime as well as at night. Various forms of lighting fitting are available, and for this purpose a fluorescent tube mounted horizontally in a parabolic reflector is particularly suitable provided the fitting is mounted in such a way in relation to the chalkboard that no undesirable reflections are created (Plates 45 and 46).

Teaching areas where small local groups are to engage in special activities can be marked out by means of selective local illumination. This is additional to the general lighting level and is, of course, in addition to that required by statutory regulation. The local lighting may take the form of special lighting units which can be lowered, where necessary, over the areas while they are being used. Such additional levels of local lighting should give at least twice and preferably three times the level of illumination provided by the general lighting if they are to create the focus of attention which is their purpose. They should, of course, give mainly direct downward lighting. Fittings which are attractive in appearance, with a touch of colour and sparkle, will probably be chosen for such positions, but care should be taken to avoid dazzle and glare. Designs should give a suitable cut-off to the bare lamp from both standing and sitting positions, and the overall luminance of the fitting should be much lower (about 100 ft-L or 1000 asb) than that of the fittings used for the general lighting.

In areas of the school which are not strictly teaching areas and which are not therefore subject to the building regulations, considerably more freedom can be exercised in the choice of lighting. Staff rooms and social areas, for example, can be given lighting of a special character, cheerful and colourful, domestic and relaxed, as a contrast with the more workaday teaching rooms. Circulation areas, dining-rooms, and assembly halls may offer a contrast in lighting character which, if well done, can be most welcome.

P.S.A.L.I. IN SCHOOLS

Most schools will work under natural lighting during daylight hours. Difficult problems of building design arise, particularly in secondary schools, where large deep rooms, such as teaching laboratories, may have to be provided in multi-storey buildings. In order to achieve full daylight penetration to the statutory level, in such rooms, very high ceilings are necessary. A solution to the problem is the use of integrated daylighting and permanent supplementary artificial lighting, and the building regulations may in due course be changed to permit such a solution.

Special dispensations are at present granted to approved schemes of supplementary lighting.

In school buildings the need for supplementary lighting is felt particularly in specialist rooms where otherwise the only means of achieving the necessary standards of daylighting would be by the introduction of top lighting which is only feasible in single-storey buildings or on the top floors of multi-storey buildings.

Permanent supplementary artificial lighting in schools can either be in the form of a distributed installation with the light arranged inset in the ceiling to give a downward component to supplement the natural daylighting, or it can be in the form of an artificial laylight placed away from the windows. The provision of supplementary lighting also permits the use of smaller windows or some other expedient to reduce sky glare because the windows alone are no longer required to provide depth of daylight penetration. An integrated system of daylight and artificial lighting can therefore create a more favourable state of adaptation for the eyes with a reduction in glare and a general sensation of higher brightness and better illumination. The general problem of the design of supplementary lighting is treated in Chapter 6.

Supplementary lighting in schools should employ fluorescent tubes exclusively. The type of fluorescent lamp to be used should be either one of the newer forms of 'colour matching artificial daylight' designed deliberately to match daylight both in colour appearance and colour rendering, or, where economy in running costs and in capital costs is more important than good colour rendering, the improved 'daylight type' of lamp of correlated colour temperature around 4500°K can be used, the cooler types of lamp being more suitable for blending with north daylight, while the warmer types are more suitable where sunlight can be expected. An artificial supplement can often be set in the ceiling and can have exactly the same effect as natural top lighting. An artificial laylight should stop short of the wall to avoid patterns of light on the walls which can sometimes be a source of distraction. Simple white-painted vertical diffusing louvers can be used to prevent direct view of the lamps while opaque light grey louvers reduce the visible brightness of the ceiling and are therefore indicated in order to reduce glare.

The supplementary lighting should be provided bearing in mind that the criterion is always that a natural appearance and a satisfactory balance of brightness should be given to the room. Reference should be made to Chapter 6.

THE LIGHTING OF AUXILIARY SPACES IN SCHOOL BUILDINGS

So far attention has been given to the main teaching areas, but there are lighting problems in other spaces in school buildings which require a brief mention. Circulation areas and corridors should be well lit, avoiding glare, and particularly avoiding the view of bright sky at the end of a passage or at the top of a stairway. Such a view of sky will not only cause discomfort from glare, but could be actually dangerous because of the dazzling effect. Lighting in circulation areas should be of not less than half the level of that in classrooms, so that they do not appear gloomy by comparison.

KITCHENS

Lighting in school kitchens is the subject of statutory regulations. The levels of illumination in kitchens must be at least the same as those in teaching areas in order to ensure that there is adequate light for staff not only to go about their work but to induce cleanliness and good maintenance. The same design requirements that apply to classrooms apply also to kitchens with special emphasis being placed on simplicity of design of the artificial lighting to ensure easy cleaning at frequent intervals.

GYMNASIA

One of the most difficult lighting problems in schools is that of gymnasia and games areas. Contradictory requirements have to be resolved. High levels of lighting are necessary in order to

Plate 44 (a, above). The original Percon fitting designed to meet the Regulations while maintaining a high light output ratio (88%). It is moulded out of acrylic plastic sheet (040 Perspex). (Photograph by courtesy of the Building Research Station. Crown Copyright Reserved.) *(b, below)*. Another simple range of fittings for tungsten filament lamps meeting the Regulations. On the left: 8 in. diameter fitting for 150/200 watt lamp; maximum brightness (200 watt) 795 ft-L (8500 asb). On the right 6 in. diameter fittings for 60/100 watt lamp; maximum brightness (100 watt) 725 ft-L (7800 asb). (By courtesy of Plus Lighting Ltd. Photograph by Dennis Hooker.)

Plate 45. The deep lining to this rooflight is painted light grey (B.S. 9–099) on surfaces facing the room to reduce the brightness contrast with the white ceiling. Simple lighting fittings using tungsten reflector lamps replace the daylight after dark in helping to light the chalkboard and display wall. (Architects: Development Group, Dept. of Education and Science. Photograph by John R. Pantlin.)

Plate 46. This directional 'wall-washing' fitting has a parabolic reflector in polished aluminium and adds about 25 1m/ft² (260 lux) when mounted 3 ft. (0·9 m.) from the wall. It can be positioned to give a sharp cut-off at the top of a chalkboard. (Photograph by courtesy of Atlas Lighting Ltd.)

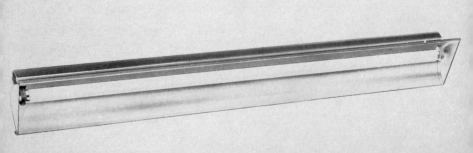

Plate 47. In this example of a school for partially-sighted children built in Birmingham 12 years ago, the windows and roof lights are supplemented by an illuminated ceiling. While this provides a high level of illumination, the brightness of the plastic diffuser is higher than is now recommended. (Photograph Crown Copyright Reserved.)

Plate 48. A school for partially-sighted children at Exeter. To help partially-sighted children to see adequately, the designer must provide a very good standard of illumination preferably with emphatic modelling, to give contrast and stereoscopic effect. In this school at Exeter, the section of the classrooms has been designed to give not less than 3% daylight factor at the back of the room. Recessed fluorescent fittings and some pendant tungsten fittings can be used to supplement the daylighting when required. Children needing very high illumination for particular tasks can take their work into a low bay near a window, or to a porch outside. Blinds are provided on all windows to control sun and sky brightness. (Architects: Stillman and Eastwick-Field. Photograph by Henk Snoek.)

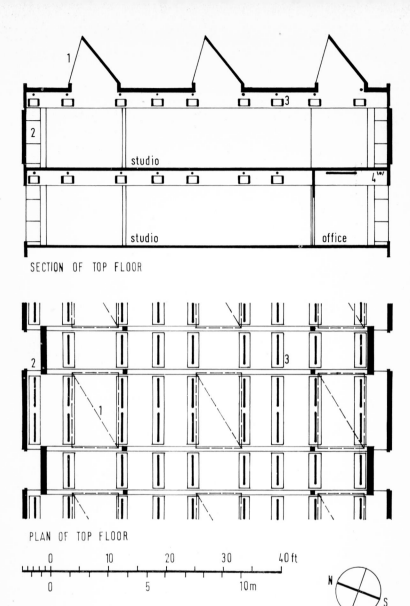

SECTION OF TOP FLOOR

PLAN OF TOP FLOOR

```
0        10        20        30       40 ft
0               5               10 m
```

Plate 49 (a, b and c). School of Architecture, Strathclyde University. The structural beams and acoustic baffles have served a double purpose as louvres for the fluorescent lighting. The artificial lighting has been designed to be used in conjunction with the daylighting, as well as after dark, and has colour-matching fluorescent tubes (Philips 55). The windows provide a view out but their size has been reduced to balance the brightness of the artificial lighting, which by itself gives 25–30 $1m/ft^2$ (260–320 lux). (Architect: F. Fielden; lighting designers: Engineering Design Consultants. Photographs by Henk Snoek.)

(*a*) 1. Rooflights. 2. Windows. 3. Fluorescent fittings above acoustic baffles. 4. Fluorescent fittings with clip-on diffusers in small offices (pendant metal tungsten fittings are also used in these offices for local lighting.

example to the exit in a corridor, while the phototropic properties of light can be used for the same purpose.

LIGHTING FOR DEAF CHILDREN

Special attention should be given to the lighting in schools for the deaf. The use of lip-reading in the teaching of the deaf or partially deaf requires that the slight movements of the teacher's lips should be easily seen from any part of the classroom. This can be achieved provided the teacher's face receives preferential lighting to a level at least three times that of the general lighting in the room, and provided the direction of the light gives good modelling of the lips, while at the same time permitting a full view of the whole expression on the teacher's face. A side window is best, with the teacher half facing the window so that light comes at an incidence of about 45° both sideways and downwards. Above all the teacher must not back the window. At night the placing of a light in the same position as the window by day will be a useful aid, but ideally the lighting should be designed with the skills of the stage-lighting designer to give the best effect, always bearing in mind that the result must be comfortable for the teacher and must not introduce glare discomfort. Equally good modelling can be obtained from a large light source of low brightness as from a brilliant spotlight, and the glare discomfort from the large source will be much less.

SCHOOLS FOR PARTIALLY-SIGHTED CHILDREN*

The lighting of schools for partially-sighted children requires special consideration. The educational trends are towards integrating such children into the general school community at the earliest opportunity because experience has shown that, once they have been taught the basic educational skills in a special

*See plates 47 and 48.

school for the partially sighted, many handicapped children can then take their place in a normal school and keep up with their fellows provided they are given some help, such as permitting them to come up to the chalkboard to check on something they have missed, or allowing them to sit near the window or in any other place they find best.

The principles of good lighting are the same as in normal schools but particular care has to be given in the application of these principles because it is usual for children with a wide range of visual disabilities to be brought together and taught in the same school and in the same classrooms. Ideally these different visual disabilities require different lighting treatment. This is one of the major problems in the lighting of such schools. In addition to enabling the children to use efficiently those visual abilities which they possess, the main aim in the design of a school for partially-sighted children is to give them visual surroundings in which the effects of their disabilities are minimized and which enable them to develop their self-confidence and self-reliance.

The various local education authorities handle the problem of the relatively small number of partially-sighted children in different ways. Some authorities may set up boarding schools for children who have to come from considerable distances, or, at the other end of the scale, in some areas it may be possible to equip a special partially-sighted unit in a normal school. Many educationists are firmly of the opinion that by far the better way to teach children who are visually handicapped is to provide special teaching facilities in normal schools so that the children can take part in normal school life as far as their handicap permits them, in order that they should not feel themselves segregated from the community because of their disability.

Teaching groups in any of these different types of school are invariably small, fifteen or less, but each child needs plenty of space and each child needs individual attention. Classrooms are therefore of about the same size as those in normal schools, but more space is left to permit the children to move about freely and with fewer obstacles in their path.

Lighting in a school for partially-sighted children should be sufficient to enable them to make the fullest use of the visual environment in their education. In addition to this, care should be taken to ensure that, both in the classroom and in the associated circulation space, they can find their way about with the minimum of difficulty. This requires the accentuating of doors, steps, handles, and rails, both by the use of light and by the use of strong and contrasting colour where this helps rather than confuses.

The chief difficulty in prescribing the lighting for such schools is that there are, in the same classroom, children who are suffering from high degrees of myopia, the common visual handicap in children, and other children who are disabled for other reasons such as juvenile cataract and nystagmus. While for all of these plenty of light *on the work* is a help, those with cataract are handicapped by the presence of bright light sources elsewhere in the visual field because they add to the veiling which is their main handicap. These problems have been investigated in some detail by ophthalmologists and lighting specialists working together, and the most workable solution has proved to be to provide a high overall level of lighting either by natural or artificial lighting or by both combined, with the maximum regard for the elimination of excessive brightness in the design, and then to allow the child to choose the position in the room which suits him best. Some children, particularly albinos, will not like the high levels of lighting which are desirable for others, and they may benefit from using the same tinted glasses which they already use out of doors. The Joint Committee on Lighting and Vision of the Medical Research Council and the Building Research Board prepared a memorandum on the lighting of schools for partially-sighted children, which, though not having statutory force, gives a useful guide, although present knowledge would suggest that insufficient emphasis was paid to the need to reduce high brightness in areas away from the work.

A level of working illumination, that is, on the desks and wherever children will be required to undertake visual tasks, must be provided, not less than 500 lux (50 lm/ft²) and preferably of the order of 750 to 1000 lux (75 to 100 lm/ft²). While levels of this order are of little advantage to normally-sighted children, since they result only in a relatively small improvement in visual performance, among partially-sighted children such high level lighting does assist them to pick out detail and to learn visually by the intelligent interpretation of visual information which to the normally sighted would be considered inadequate. The small improvements which result from levels of lighting above the normal may bring a child across the threshold of vision and give him confidence to comprehend the world around him. If he can gain this confidence in school, he can often maintain it in the world outside.

In order to provide this high level of illumination, natural lighting will usually have to be supplemented with permanent artificial lighting. One satisfactory method of integrating artificial lighting and daylighting in such special schools is to employ a luminous ceiling over a substantial part of the room. This may consist of a small-scale louvered grid which screens the bare lamps from any direction of view except looking directly upwards. The average luminance of such a luminous ceiling should be not more than 150 ft-L (1500 asb).

Screened ceiling lighting of this kind can, of course, be provided in a single-storey school by a top light in the roof. The combination of top lighting, whether by daylight or by artificial light, with side windows is particularly satisfactory because it enables high levels of working illumination to be provided while still ensuring that a view outside is possible without the introduction of excessive sky glare from the upper parts of side windows. Side windows must never become sources of glare. Their size, height, and position should be chosen so that views of large areas of sky by the children are avoided, and in particular blinds which are both optically efficient and mechanically reliable should be installed and maintained in good working condition so that they can be used easily by the staff. Consideration should also be given to the control of the brightness of the scene

outside the window by the use of overhanging eaves, projecting walls, and well placed trees and shrubs.

It is most important to avoid the use of specular reflecting
materials which cause reflections which could dazzle the child
and so swamp what he is trying to see. Desks and table tops, wall
surfaces, and floor finishes should be matt and light in colour.
Any use of specular reflecting materials should be confined to
those which are used deliberately for a special reason.

Additional directional lighting will be required over and above
the general lighting, no matter how high the level of this general
lighting. Handicrafts and other activities of special visual difficulty will benefit from directional lighting to create highlights
and shadows to model the objects in a way which allows them to
be identified and handled with ease. Some of this directional
lighting is quite simply provided by arranging that this kind of
handicraft work can be done at the window. In other cases local
lights, either directional from the ceiling or the walls, but out of
harm's way, can be used to supply the modelling lighting.

Any demonstration areas to which the children's attention is
to be drawn should also have additional local lighting. Relatively
little use may be made of chalkboard instruction because
many of the children may be unable to derive any benefit. Suggestions have been put forward for the use of self-illuminated
chalkboards, that is, a board consisting of a sheet of white
diffusing plastic behind which are placed a bank of fluorescent
lamps. Writing is done with a thick black crayon. Such devices
are often glaring and uncomfortable to use for long periods.
Other optical aids such as portable magnifiers with or without
auxiliary lights have been produced and in some cases have
been used in schools. Experience has shown, however, that in
a well designed classroom with good lighting, children often
prefer to dispense with such troublesome optical aids because
the advantages which they give are so marginal as not to compensate for their mechanical inconvenience. Many of these aids
are not truly related to the kind of visual world in which these
children live. Those children who are highly intelligent can often

develop powers of information interpretation which are quite remarkable. Optical aids which are designed from the concept that the eye is a camera, and in the case of a partially-sighted child an inefficient camera, fail to take into account that sight is a combination of optics and interpretation and that often, by training the interpretation the defects of the optical system can be better overcome than by providing mechanically inconvenient visual aids, which in any event are unlikely to be available in the outside world in later life.

Provision should also be made to permit instruction to take place out-of-doors so that the full use of full sky light and sunlight can be made. In Great Britain the weather is not often favourable for such outside teaching, but the provision of glazed windbreaks and a glass roof can give the lighting advantages of a greenhouse without its unwanted heating disadvantages.

SPECIAL TEACHING MATERIAL FOR THE VISUALLY HANDICAPPED

Lighting can help considerably in the teaching of partially-sighted children, but much can be done by improving the visual teaching material. Unfortunately, children with different handicaps need different aids. Children with high myopia can see quite well if they hold their work close to the eyes, but they cannot read easily from a chalkboard, and so for them the main additional need is for very large detail in distant material. Few of them are likely to want books with large print, which are clumsy to use.

Other children will, however, benefit from such books. These are, however, usually printed with black lettering on white paper. Some children with cataract, however, see better with white lettering on a dark background because the scattered light from a white page veils the faint image of the letters. 'Negative-printed' reading material can be prepared photographically and used to test whether a child with cataract will benefit from its use.

8

The Lighting of Hospitals

There are three main components in a hospital building complex which require special and individual lighting treatment. The first comprises the spaces occupied by the patients, the sick wards, the day spaces and visiting rooms; the second includes operating theatres, recovery rooms, diagnostic centres, and examination rooms; while the third group includes laboratories, service areas, kitchens, sterilizing, and laundry rooms together with the general administrative offices and staff rooms.

The visual requirements in the different parts of the hospital must each be examined individually, because the application of the basic principles of good lighting is different in each case.

WARDS AND DAY SPACES

The lighting of hospital wards has been given particular attention during recent years, particularly as the result of the development, at the Building Research Station, of methods for the evaluation of the subjective aspects of lighting quality which are of overriding importance in hospital wards. The psychological problems arise from the fact that in any given ward there may be patients in different degrees of sickness, from those who are critically ill to those who are on the point of leaving for home. In a large hospital, of course, it is easier to segregate people in different stages of illness, but in a small hospital it becomes essential to provide lighting which, while not too restrictive for convalescent patients, is nevertheless comfortable for those who are less well.

Lighting in hospital wards has also been closely linked to the development of a different form of layout of beds and service areas. The traditional 'Nightingale' ward with perhaps forty beds ranged on either side of a central walking space has been superseded by wards divided into small units with not more than six beds, arranged three a side. In addition, the growing practice of taking patients to a separate examination room has obviated the need for examination lighting to be provided in the general ward.

The main requirements for general ward lighting are a sufficient level of general lighting to enable the nursing staff to go about their work efficiently and to be able to survey the ward and spot instantly anything that is amiss, and sufficient light to enable the patient to be able to see to read, sew, and engage in any other activity without disturbing another patient who may wish to rest. In addition, there must be an absence of any form of glare or visual distraction or irritation (Fig. 17).

The design of ward lighting is to some extent a compromise between sufficient light for the medical staff and complete freedom from any excessive amount of light which might disturb sick patients. The compromise is easier to achieve when clinical examination is not undertaken in the ward itself, because then the visual tasks which the medical and nursing staff have to perform in the wards are not very exacting. The nurse must, of course, be able to see immediately if anything is wrong with the patient, without having to rely upon any auxiliary lighting, and the level of general lighting in the ward must be sufficient for this to be done. Apart from this, however, provided the lighting in the ward is sufficient to enable the nurse to take immediate action if anything is wrong with the patient, the general lighting can be of quite a low level, well below the discomfort threshold for any but the most distressed patients.

An investigation of the ward lighting problem undertaken at the Building Research Station has led to recommendations from the Ministry of Health, which at the time of writing do not have statutory force, but which give clear directions both as regards

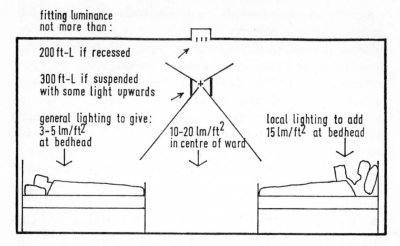

fitting luminance
not more than:

200 ft-L if recessed

300 ft-L if suspended
with some light upwards

general lighting to give:
3–5 lm/ft²
at bedhead

10–20 lm/ft²
in centre of ward

local lighting to add
15 lm/ft² at bedhead

Fig. 17. Principles of Hospital Ward Lighting. The Ministry of Health recommends that ward lighting should be concerned primarily with the comfort of the patient in bed, whether lying down or sitting up. This can be achieved by planning the ward layout so that no patient has to face directly toward a window and by designing the windows to avoid glare. The artificial lighting installation should have a Glare Index of not more than 13, using fittings for the general ward lighting designed so that the patient cannot see parts brighter than 300 ft-L (3200 asb) from normal positions in bed. This general lighting should be supplemented by bedhead fittings under the control of the patient and with the degree of adjustment limited so that other patients cannot be disturbed by its use. In addition, dim watch lighting is required at night.

the quantity of light to be provided in wards and also on precautions to be taken to avoid excessive glare, discomfort, and distraction. In particular these recommendations suggest that the general level of illumination in the ward should be quite moderate, of the order of 30–100 lux (3 to 10 lm/ft²) but that there should be lighting at each bed head under the control of the patient to enable him to provide himself with additional light should he wish to read or work.

In terms of daylighting, the Ministry of Health recommends that the minimum daylight factor in a ward should be not less than 1%. In addition, the sky should be screened from patients near the windows to prevent them suffering from sky glare on bright days.

In order to achieve comfortable daylighting conditions, the windows must be designed with considerable care. Some form of permanent glare protection device should be built into the design in order to limit the amount of visible sky to those occupying the beds nearest the window. While in many buildings adjustable blinds of the venetian type are quite satisfactory, in a hospital the staff do not have the time to lower and adjust these blinds, and patients, of course, are not able to do this adjustment themselves. One of the most useful methods for achieving adequate penetration of daylight into the depth of a four- or six-bed ward is the use of a horizontal baffle so placed in the window that little or no obstruction results to daylight penetration yet almost complete obscuration of the bright upper part of the sky is achieved for the patient nearest the window.

This type of window design will only be successful if the interior surfaces of the ward have high reflectance, because the system depends upon the maximum inter-reflection of light from all the surfaces in the room. The average reflectance of all surfaces in the room, walls, ceiling, floor, internal furnishings, and hangings, should be of the order of 40%, and certainly not less than 30%. If this is not ensured, the lighting will appear inadequate in the deeper parts of the room.

With a window design of this kind, it is relatively easy to meet the Ministry of Health requirements for minimum daylight factor of 1% in a ward in which no bed is more than 15 ft. from the window. In deeper wards, it may be desirable to use some form of permanent supplementary artificial lighting in the deeper parts of the room. This, of course, does not meet the Ministry recommendations, but since these recommendations do not have statutory force, it may well be possible to convince the authorities that the use of a permanent artificial supplement, provided this is well integrated with the daylighting design, may be an acceptable solution to a difficult lighting problem. The use of permanent supplementary artificial lighting does not remove the need to design the windows with a baffle or louvre system to reduce sky glare to the patients near the windows. The supplementary light has its effect only in the parts of the room remote from the windows. Such supplementary lighting, which should be provided by fluorescent lamps, should preferably be built into the ceiling, and integrated with the structure to provide a clean, hygienic and easily maintained system. The system should ensure that the lamps themselves are not visible from the beds, bearing in mind that recumbent patients can look directly upwards and hence the conventional types of vertical louvering are not acceptable if the lighting unit is placed above a bed. The supplementary system should be mounted so that it is not too obtrusive during the daytime.

WARD LIGHTING IN THE EVENING

There will be long periods, especially in winter, when the hospital ward is in activity although it may be dark outside. These periods will be from waking time onwards until daylight has increased to a sufficient level, and from dusk until 'lights out'. During this period the artificial lighting will create the visual environment.

At night-time many people prefer filament lighting because it is more pleasant, warmer, and less institutional. Filament lighting is associated with the home, while fluorescent lighting is associated in people's minds with work places and official buildings. With skill, however, fluorescent lighting can be designed for use in wards without introducing the unwanted institutional character.

Following investigations at the Building Research Station, the Ministry of Health have produced suggestions for the manner in which wards should be lit over and above the recommendations for the level of lighting. These recommendations have particular importance because for the first time emphasis has been placed officially on freedom from glare and on the absence of distressing brightness from the position of the patient, and on the need for a comfortable and pleasant environment in a hospital ward, rather than as before on the provision of high levels of illumination.

The distribution of light should give a pleasant and warm character to the ward, but the general effect should not be dull and soporific. Totally indirect lighting to a low level of illumination is soporific and annoying because the eyes are constantly being drawn up to the bright ceiling. Equally, lighting which employs small, brilliant sources of light, though stimulating and cheerful, can be irritating if overdone. The light distribution worked out by experiment at the Building Research Station permits about 60 % of the total light to go upwards to be reflected back from the ceiling and upper walls. In addition, direct light is permitted on the central circulating space in the ward and on the lower parts of the bed, but no direct light should ever reach the patients' eyes. The recommended light distribution is therefor a combination of indirect light (the greater proportion) and of direct light, of which the patient himself receives only indirect light.

Lighting fittings to meet these requirements of light distribution were designed at the Building Research Station both for filament lamps and for fluorescent lamps. The Ministry of Health recommendations are that no part of any lighting fitting as seen from any normal primary line of sight, including that of a reclining patient, should exceed 300 ft-L (3000 asb). The Building Research Station fittings keep the maximum luminance well below this figure in order to ensure that there is no possibility of glare arising in any circumstances. In addition to the luminance limits incorporated in the Ministry of Health recommendations, there is the additional recommendation that the Glare Index in the ward should not exceed 13 as calculated for the view of any patient in any usual or necessary position. The Building Research Station design of fittings ensures that this Glare Index limit will not be exceeded.

This general ward lighting must be supplemented by bed head lights. These should be mounted over each bed and should be controllable by the patient in such a way that he himself receives fully adequate illumination, possibly up to a level of 1000 lux

(100 lm/ft²) but in any case not less than 150 lux (15 lm/ft²), this illumination to be provided on his book whether he is sitting or lying down and where it is necessary for the fitting to be adjusted to provide the light in the position required, it should be easily controllable by the patient. The patient has also to be able to switch the light off when he does not require it, but the degree of control which he has must not be sufficient to enable him to tilt the fitting so that the light shines directly in another's eyes (Plate 54 a and b).

In hospitals where no separate examination room is provided, the adjustable bedside light may also have to meet the clinician's requirements. This requires a greater degree of adjustment and it is therefore recommended that such bed head lighting should have some form of built in lock which can be released by the clinician to give him the extra adjustment which he requires and the extra level of illumination which he may need.

The Ministry of Health recommendations do not endorse the use of wall-mounted fittings which send light upwards for the general ward lighting and which also have a downward component for reading. It is difficult to provide satisfactory lighting of this kind. At its worst, bright patches of light distract the patient's view wherever he looks.

NIGHT-TIME LIGHTING IN WARDS

During the night while patients are sleeping, there must be sufficient light in the wards to enable the night nurse to see immediately if anything is wrong. At the same time the lighting must not be sufficiently bright to prevent sick patients from sleeping.

The Ministry of Health recommendations are for a level of illumination in wards at night-time of the order of 0·01 lm/ft² (0·1 lux). This is comparable with full moonlight and is sufficient to enable the night nurse to see sufficiently well once her eyes are adapted to the lighting in the ward. The eyes of young people, such as nursing staff, can adapt to this level of lighting within a few seconds after coming from a moderately lit service room or corridor.

The night-time lighting must be provided from sources which are completely concealed from any possible view of the patients. Such night-time lighting is best provided from units suspended from the ceiling or mounted in the ceiling, rather than from units mounted in the floor. There is some argument for the night-time lighting being provided from a fluorescent source because it has a 'cool' appearance which is apparently beneficial to feverish patients.

FLUORESCENT AND FILAMENT LIGHTING

For many years there was considerable prejudice, particularly on the part of clinical staff, against the use of fluorescent lighting in hospitals. These prejudices were justified in the early days of fluorescent lighting, because the colour rendering, particularly of the complexions of sick people, was unsatisfactory. In addition, the engineering of the early lighting installations was bad, and employed industrial lighting fittings totally unsuited to the character of a hospital ward. The Joint Committee of Lighting and Vision of the Medical Research Council and the Building Research Board sponsored an investigation into the use of fluorescent lighting in hospitals, which resulted in the interesting conclusion that clinical staff actually preferred a certain form of fluorescent lighting for the various critical judgements of colour which they are called upon to perform in the course of their duties. The experiment was so arranged that when making these judgements, the clinical observers were not able to see what type of lighting was provided. In this way any prejudices which they might have had against fluorescent lighting did not play a part in their judgement.

The form of fluorescent lighting which was shown to be superior in every way (except one) was a type of fluorescent lamp with a correlated colour temperature of the order of 4000 °K, and with a full content of red radiation (in which fluorescent lamps are normally deficient). As a result of these experiments, the Ministry of Health adopted in principle the recommendation

LARKFIELD HOSPITAL, GREENOCK

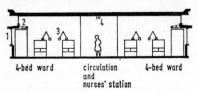

4-bed ward   circulation
and
nurses' station   4-bed ward

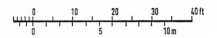

WEST MIDDLESEX HOSPITAL

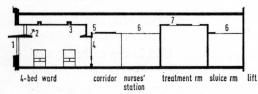

4-bed ward   corridor   nurses'
station   treatment rm   sluice rm   lift   treatment rm   corridor   4-bed ward

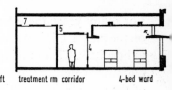

Fig. 18. Windows with 'eyebrow' transom to reduce glare for patients near the window. Minimum daylight factor 2%. Average reflection factors: wall 60%, ceiling 80%, floor 40%.
1. Window (part has glass to floor)
2. Tungsten fittings concealed i transom light the ceiling an window wall
3. Adjustable bedhead fitting
4. Open bottomed tungsten fittings i circulation. At night the nurse station has desk lights, and there dim watch lighting from low lev fittings

Fig. 19 (a and b). In this building 100 ft. deep, the wards are daylit to a 2% minimum daylight factor, using the same type of 'eyebrow' transom as in Fig. 18. The adjacent circulation and nurses' stations are lit by a combination of daylight and P.S.A.L.I., while the internal workrooms have permanent artificial lighting. (Architects: Robert Matthew, Johnson-Marshall and Partners.)
1. Double glazed window (the hospital is near to London Airport)
2. Fluorescent indirect general ward lighting
3. Tungsten bedhead fittings recessed in ceiling
4. Glazed screens with adjustable blinds
5. Fluorescent lighting above diffusing 'sandwich' of two P.V.C. skins enclosing louvers (Ionlite), linked to a time switch giving higher levels during the day
6. Fluorescent lighting above diffusing 'sandwich' faced on underside with impregnated paper skins with sound-absorbing properties
7. Recessed fluorescent troffer fitting with Perspex diffuser

Levels of artificial illumination after dark are:

Wards 15–20 lm/ft² (160–210 lux)

Offices 30–35 lm/ft² (320–365 lux)

Corridors and internal workrooms

     60–65 lm/ft² (640–700 lux)   } according to nature
     40–45 lm/ft² (430–480 lux)   } of work done

(see also Plate 52 opposite)

Plate 52. West Middlesex Hospital. *(a, above)* View of four-bed ward. *(b, below)* Corridor and nurses' work station (see Fig. 19 *above*).

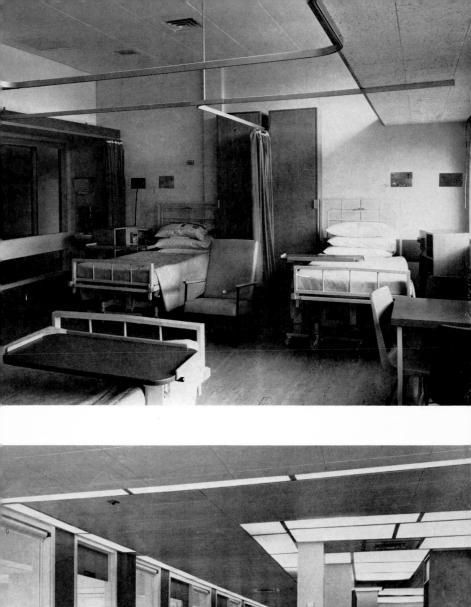

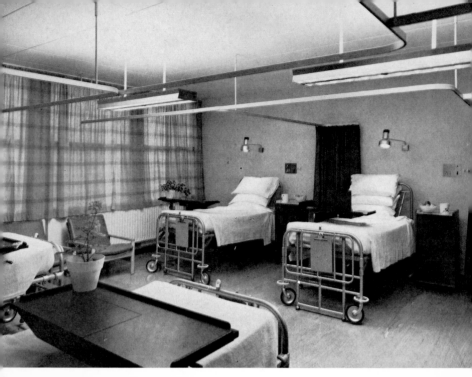

Plate 53. In the wards of St. Mary's Hospital, Portsmouth, the fluorescent fittings give 15 1m/ft² (160 lux) between the beds and can be supplemented by the incandescent bedhead fittings. Both types of fittings include 15 watt lamps which can be used for watch lighting at night. The Glare Index is within the required limits but the ceiling may appear rather bright in relation to other surfaces in the room. (Photograph by courtesy of Atlas Lighting Ltd. by Studio Jaanus Ltd.)

Plate 54 (a, left). Bedhead fitting for 40/60 watt incandescent lamp (with 15 watt pigmy night light). The internal reflector surface is of low brightness, and the vertical adjustment is restricted. (Photograph by courtesy of Atlas Lighting Ltd. by Shrivell.) *(b, right).* This fitting (used in the hospital shown in Fig. 19) is recessed into the ceiling above the bed. It gives an illumination of 17·5 1m/ft² (188 lux) at the bedhead, and has a low brightness from other parts of the ward. The specular louvre is rated BZ1. (Photograph by courtesy of Frederick Thomas & Co. Ltd.)

Plate 55. These tungsten fittings in New Zealand House were designed by the architects to give about 15 lm/ft² (160 lux) and were intended to be used in combination with desk lighting. The client has preferred not to use desk lighting, and the ceiling mounted fittings have been altered to emit more light with a plastic diffuser in place of the lower metal drum. (Architects: Robert Matthew, Johnson-Marshall and Partners. Photograph by Henk Snoek.)

Plate 56. In the offices of Birds Eye Foods Ltd., the artificial lighting is used throughout the day in all parts of the building. Lines of half-round reflectors, enamelled white, are recessed into the metal acoustic ceiling on a 5 ft. 3 in. (1·6m) square grid, and 5 ft. 50 watt 1 in. diam. fluorescent lamps give an illumination of 32 lm/ft² (330 lux). There is continuous fixed glazing, and mechanical ventilation is incorporated alongside the light fittings. (Architects: Sir John Burnet, Tait and Partners. Photograph by R. Gordon Taylor.)

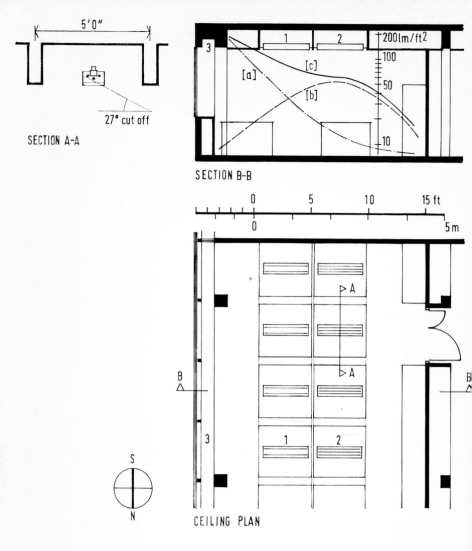

5'0"

SECTION A-A

27° cut off

SECTION B-B

200 lm/ft²

[a] [c] [b]

0   5   10   15 ft

0   5 m

S N

CEILING PLAN

Plate 57 (a and b, opposite). In the offices of the Hemel Hempstead Town Hall the lighting is a planned combination of daylighting and P.S.A.L.I. The large windows (approx. 25% of the floor area) give a high level of illumination on the desks by the windows but the room looks onto a central courtyard and the illumination falls off steeply to a daylight factor of 0·5% at the back (curve (a) on Section B-B). Supplementary lighting to a maximum of 45–50 lm/ft² (480–540 lux) (curve (b)) is provided by all three lamps in the inner row of fittings (2) which are recessed and louvred but throw light up into the ceiling cavity to reduce brightness contrasts (Section A-A). The combined horizontal illumination is shown in curve (c) in Section B-B. After dark two lamps in each row (1 and 2) provide illumination over the main working area. (Architects: Clifford Culpin and Partners.)

Plate 58 (below). This production laboratory required a controlled environment and is lit by artificial lighting alone to 30–35 lm/ft² (320–370 lux) with additional local lighting where required. The ceiling dominates the interior with banks of louvred fittings, high colouring, and strongly modelled, sound absorbent pyramids (the process is a noisy one). Another approach to windowless interiors that has been tried is to incorporate laylights in walls or ceilings as substitute windows. Perhaps the best method is to treat the lighting as if it were a normal scheme for use after dark, with particular attention to quality and variety. (Designers of the County Laboratories: Jonathan Green Associates. Photograph by Ian Macdonald.)

Plate 59. This factory for Rembrandt Dresses Ltd., uses the 120 ft. span double-T concrete roof beams as large-scale louvres to give a partial cut-off to the bare fluorescent lamps which provide the working illumination throughout the day. Not more than five rows of lamps can be seen from any position on the factory floor, and glare is kept to a tolerable level as the lamps are seen against a bright background. The level of illumination on the sewing machines is 70 lm/ft^2 (750 lux). Although the working illumination is provided by the artificial lighting, the variety and modelling from the view windows can be appreciated from all parts of the production floor (the photographer had his back close to a window). Some landscaping has been done to improve the view out. (Architect: W. H. Marmorek. Photograph by courtesy of Atlas Lighting Ltd. by Studio Jaanus Ltd.)

Plate 60. Local lighting giving 200 lm/ft^2 (2150 lux) for intricate electrical assembly work at Reid and Sigrist, Ltd. (Photograph by courtesy of Crompton Parkinson Ltd. by A. C. K. Ware (Photographs) Ltd.)

that hospitals should be lit by fluorescent lighting of this particular type, particularly in the areas, including the wards, where clinical judgements might be required. A full specification in terms of correlated colour temperature and colour rendering characteristics was prepared by the Ministry to be the subject of a recommendation.

The one exception to the overall satisfactory character of this form of fluorescent lighting is the recognition of jaundice. In early jaundice, the patient's colour changes and takes on a slightly yellowish cast. This change of colour is readily recognised in the bluish light of the clear sky or of the overcast sky. But fluorescent light of the Ministry of Health specification is a little too biased away from the blue region of the spectrum to be fully satisfactory. The 'daylight' type of fluorescent lamp, which is otherwise too cold and lacking in red radiation, is excellent for the recognition of early jaundice, but since this is the only criterion by which it is superior, its use is not indicated for general lighting. The Ministry have therefore coupled their recommendation of fluorescent lighting at 4000°K with a suggestion that special provision should be made where necessary and possible for special lighting where early jaundice may be suspected.

Fluorescent lighting, even of the recommended type, renders colours and particularly the colours of the patient's skin in a somewhat different way from daylight. This need not be disconcerting provided that clinical staff become thoroughly familiar with its use in all parts of the hospital where they are going to have to make judgements of the condition of the patient by his skin colour. Anaesthetists would normally use the skin colour as one of their more reliable criteria. Medical staff are now coming round to the opinion that, provided the whole of the hospital, and preferably all hospitals everywhere, are lit by exactly the same type of good quality fluorescent lighting throughout, familiarity and adjustment to this type of lighting would enable them quickly to learn to adjust their judgement of skin coloration to the colour rendering given by the lighting. Their main difficulty is in passing from a ward lit by one form of lighting to another

14

room lit by another form of lighting because then the skin colour changes and judgement is upset.

THE LIGHTING OF OPERATING THEATRES

The main requirements of operating theatre lighting are to provide a very high level of illumination of the order of 20,000 lux (2000 lm/ft²) over the cavity area, as free as possible from unwanted shadowing but still retaining modelling and depth revelation, together with facilities for adjusting the position and direction of the source of light and to permit control of this direction quickly and without difficulty. In addition, the brightness in the room itself must be such that all the other occupants, in addition to the surgeon, are working in visual comfort and can see to go about their jobs efficiently without in any way interfering with the lighting employed by the surgeon himself.

Of the various lighting systems in general use at the present time, the great majority consist of local lighting systems which are in the direct control of the surgeon himself or of his assistant. Less frequently used but very strongly advocated by their supporters are systems which employ external parabolic mirror domes, or adjustable and movable projectors located outside the operating enclosure but under the control of the surgeon or his assistant.

The conventional 'operating theatre fitting' has the advantage of relative cheapness, ease of maintenance and replacement, and of being under direct mechanical control of the operating staff. Most units of this kind consist of a single light source from which the light flux is collected by a very large mirror or lens system, or both combined, to concentrate the light into a beam which can give either a fine or a coarse focus on the operating area. The older type of operating theatre fitting allowed little or no control over the focusing of the beam, and so there was no control over the size of the area which could be efficiently illuminated. More modern types of unit give a wide control, although too much complication in the optical system is inadvisable because the staff may not be able to use the system effectively.

The optical system must provide a high illumination at the cavity, and it must ensure that the amount of illumination will not be seriously affected as the surgeon moves about over and around the cavity. The illumination must therefore reach the cavity from a large area in order that the shadow cast by the surgeon will be of little or no significance. It is desirable that the beam can be concentrated where necessary into a narrow pencil of light and also that additional beams of light should be directed into different parts of the cavity from different directions in addition to the overall illumination.

External lighting systems owe their existence partly to a dislike by surgeons of having a large and cumbersome piece of equipment close beside them while they work, and partly to the need to keep the operating theatre thoroughly aseptic. Among the very many different forms of external theatre lighting that have been devised, one is the use of a vaulted ceiling containing a large number of low intensity projectors. Each individual projector is directed at the table so that not only are variations of the level of illumination possible, but groups of lamps can be brought into play to change the direction and intensity of the main and of subsidiary beams of light. Among the advantages claimed for this system is the ability to light more than one surgical area at a time as, for example, in transplantation operations. Another form of external system uses one powerful lamp only, with a number of mirror reflectors, whose position and direction is under control by the surgeon, giving him the necessary flexibility in light control which he needs.

It is evident that the more complex the optical system of the operating theatre lighting and of the electrical dimming controls or other forms of adjustment, the more skill and understanding of lighting principles is required by the operator or his assistant. It has been proposed, particularly in France, that in view of the complexity which has to be built into modern operating theatre lighting equipment, it may be necessary for one of the assistants at an operation to be a lighting specialist who can control the lighting system according to the needs of the surgeon and in sympathy with his actions. It would, however, seem preferable that where possible quick, accurate, and trouble-free remote control lighting by the surgeon himself should be available.

While a great deal of attention has been given to the optical, mechanical, and electrical design of operating theatre fittings, relatively little attention has been given to the almost equally important problem of the lighting of the rest of the theatre. The general lighting in the operating theatre appears to be a matter of greater controversy than the lighting of the operation area itself. While many surgeons are convinced that they work better when the rest of the theatre is dark, others feel that they are happier in a room which is itself brightly lit. Experiment at the Building Research Station established that maximum visual comfort arises when the surroundings to a difficult visual task are of only slightly lower brightness than the task itself. This finding was confirmed by experiments by Putnam in Cleveland, Ohio, who demonstrated that, when operators have control of the lighting in the surroundings to the theatre, they achieved the greatest visual comfort when the illumination in the room was of the order of 1400 lux (140 lm/ft²). While individual preferences must be recognised when considering a highly individual activity like that of a surgeon performing a difficult operation, there seem to be good grounds for recommending that provision should be made for general room lighting up to a level of the order of 1500 lux (150 lm/ft²) although it is equally important that if this general lighting is provided, control of the level should be at the command of the operating surgeon who alone can decide whether he wishes to concentrate his attention, possibly for short periods, by switching off the general lighting, or whether he may wish to work with the general lighting on. Environmental design can do no more than to advise him how the lighting should be used and to give him a sufficiently flexible installation to obtain what he needs for his work.

The general lighting in the theatre can employ fluorescent

lamps of the type recommended by the Ministry of Health. The difference in colour rendering characteristics between this type of fluorescent tube, and the high efficiency filament lamp usually employed in the main operating theatre fitting, is not so great as to cause serious error in colour judgement. It must be remembered that the anaesthetist and the surgeon are concerned with different aspects of colour rendering, the anaesthetist with the skin colour of the patient and particularly his facial colour, while the surgeon is concerned with colour rendering in the cavity itself. Fluorescent lighting has the advantage that it emits very much less heat for a given amount of light than does a filament lamp. The high proportion of heat rays in filament lighting can be troublesome and indeed dangerous, and special efforts have to be taken to dissipate it. The use of fluorescent lighting avoids many of the special problems of heat dissipation.

CLINICAL EXAMINATION ROOMS

It is now customary in all new hospital buildings to provide special rooms for the clinical examination of patients in order to avoid the necessity for this being done in the wards themselves. Such examination rooms require good general lighting to an average level of not less than 200 lux (20 lm/ft^2) together with an adjustable source of high illumination, giving not less than 1000 lux (100 lm/ft^2) which can be positioned at will by the clinician and which can be adjusted either to illuminate a large or a small area to this necessary high level. The basic problem of the lighting of examination rooms is very similar to that of the surgical operating theatre itself, although the precise demands are somewhat less exacting.

The general lighting of the room should be consistent with the general lighting of the rest of the hospital. In particular, the colour of the light source which supplies the dominant lighting under night-time conditions should be the same as that in other parts of the hospital where critical colour judgements are to be made. If fluorescent lighting is used, it must be to the specification of the Ministry of Health and it must be used throughout the hospital in all areas where clinical judgements are made.

Since it is likely that many of the patients who are taken into the examination room will be seriously ill, special effort should be made to screen all bright sources of light from the patient's possible line of sight.

The examination lighting fitting itself should have a wide range of adjustment. It should be able to illuminate a small area of, say, 1 sq. ft., while it should also have an adjustment to illuminate a much larger area, for example, an ellipse of major and minor axis 4 ft. and 2 ft. respectively. It may also be desirable to provide a small spot light for precise local examination, although this may often be provided by a portable hand torch rather than by a fixed lighting fitting.

NON-CLINICAL AREAS—
LABORATORIES AND SERVICE ROOMS

In a modern hospital there are many rooms which have specialised functions and which therefore require special attention to the lighting. The lighting of laboratories and special technical rooms must be planned in relation to the visual tasks which have to be undertaken. Most of these involve some degree of difficulty and require not only a good level of general lighting, but also some form of preferential lighting on the task itself. Many of the visual tasks in laboratories require such high levels of illumination that it would be quite unsuitable to design general lighting, whether of daylight or of artificial light, to these levels. Hospital designers have therefore been encouraged to plan their laboratories in such a way that those tasks which require the most light are located near to the windows during the daytime, while those which are not so demanding visually can be performed in other parts of the room. At night a level of general lighting of the order of 300 lux (30 lm/ft^2) should be provided throughout the laboratory but with additional adjustable light on those tasks which require it.

An analysis of the tasks which are undertaken in a hospital laboratory revealed that rather less than 10% of laboratory workers' time is taken up with tasks of great visual difficulty, for example, slide preparation and mounting, etc. These difficult tasks require local illumination of the order of 1000 to 2000 lux (100 to 200 lm/ft^2). The amount of light and its direction also should be adjustable by the worker himself. Most other laboratory tasks can be undertaken in the general lighting, but no worker should be precluded from using an adjustable light if he feels he can work better that way.

Under daylight conditions it is not advisable to attempt to provide full working light by daylight alone. Except on bright days, this would demand windows which would be uncomfortably large, or it would limit the usable laboratory space to within a few feet of the window. Hospital laboratories are particularly suitable for an integrated system of permanent supplementary artificial lighting (see Chapter 6). The permanent artificial supplement should employ fluorescent lamps, and if the rest of the hospital is also lit by fluorescent lighting, it should be one of the same spectral quality, of the specification recommended by the Ministry of Health.

The daytime and the night-time lighting in a hospital laboratory must be planned together. In laboratories, as distinct from wards, the levels and type of lighting necessary at night time are essentially the same as by day. Care should be taken to avoid glare, and in particular the Glare Index should not exceed the recommended limiting value of 19.

In the compact plan ('race-track') design of hospitals, the laboratories are located in the internal core and so are lit entirely by artificial light. If staff have to pass from the laboratory to the wards, there is an adaptation problem, particularly at night. It may then be necessary to reduce the lighting in the laboratories to a somewhat lower level in order that staff passing from the laboratory through an intervening corridor to the ward may be able to adapt quickly to the lighting level.

X-RAY DIAGNOSTIC ROOMS

Rooms used for X-ray diagnosis require special treatment. In the diagnosis room, when the radiologist is examining the screen, which has a very low brightness, the room should either be in darkness or illuminated to a very low level of the order of 0·1 lux (0·01 lm/ft²), possibly with red light. Red light is chosen because it has little or no effect on the dark adaptation process and so the radiologist can study objects in the room by this low level of red light, while looking back at the green X-ray screen without any loss in dark adaptation. The screen itself should, of course, be protected from any illumination from this red room light which would otherwise tend to affect contrast.

The brightness of the panel against which the radiographs are examined must be high enough to reveal all necessary detail but not so high that excessive glare discomfort is set up in the dark-ened room. The brightness of the screen should be controllable on dimmers, bearing in mind that the comfortable range of background brightness will be in the region of 500–2000 asb (50 to 200 ft-L). Control of the screen brightness is desirable not only because it controls the degree of detail that can be seen, but also because radiologists differ in their sensitivity to glare, some tolerating higher brightness than others.

Provision should also be made to enable clinical workers and patients to adapt visually when proceeding in and out of the diagnostic room. An ante-room may be provided in which patients can be accommodated before proceeding into the X-ray room itself. This ante-room should be illuminated to about 1 lux (0·1 lm/ft²), that is, an intermediate level between that of the rooms outside from which they have come, and the dark X-ray room itself into which they have to go. This low level of illumination may be of orange or red light to assist adaptation. It must be remembered that clinical staff having been in the X-ray room for some time, will be fully dark adapted to the low level of illumination prevailing and will be able to see obstacles etc. without difficulty, whereas the patient, unless given the opportunity to adapt, will see nothing and may well cause difficulty and annoyance by stumbling over objects which are easily apparent to those already in the room. It must also be borne in mind that some people adapt to darkness more slowly than others and that certain forms of disease affect dark adaptation.

THE LIGHTING OF CORRIDORS

Corridor lighting in hospitals has to be graded to the levels of light in the rooms which are served. After nightfall but before 'lights out' wards will be illuminated to the general level of 30–100 lux (3–10 lm/ft²) whereas service rooms and laboratories will still be illuminated to the higher level which their visual tasks demand. The corridor should therefore be lit to an intermediate level, of the order of 200 lux (20 lm/ft²). After the ward lighting has been extinguished, however, and there is only the feeble night-time lighting in operation, the lighting in the corridors should be dimmed down to a level of the order of 10 lux (1 lm/ft²). If the night nurse then finds it necessary to go into a brightly lit service room, she will have a few seconds to re-adapt before going back to her duties in the ward. The problem is more acute for elderly staff who adapt more slowly than younger people.

Apart from the question of lighting level, the main requirement of corridor lighting is that the lighting units should be so designed that they cause the minimum of glare to recumbent patients on a wheeled trolley. Indirect lighting or fully diffusing units with no possibility of view of a bare lamp are therefore indicated.

Corridor lighting requires special attention in compact plan hospitals where the wards are on the periphery of the block and the service rooms are in the internal core. Traffic between wards and internal rooms will be frequent and will always be through and across the corridor, so special attention has to be paid to adaptation problems especially at night.

LIGHTING AND INTERIOR DECORATION
—COLOUR

An integrated system of lighting, both daylighting and artificial lighting, requires that the interior surfaces of rooms must have specified reflecting characteristics in order that lighting levels can be computed with precision. In deep wards, including those characteristic of the compact plan of hospital design, adequate daylighting can only penetrate the back of such rooms provided that internal reflectances are set and maintained at a high value. Again in order to obtain the desirable character of the artificial lighting for use in the evening in wards, the reflectances of the internal surfaces must be related to the light distribution in order to provide the necessary brightness distribution.

It is therefore clear that the reflecting characteristics and therefore the interior decorating scheme in a hospital is as much a part of the lighting design as are the windows and the lighting fittings. It is essential that any scheme of redecoration and maintenance reproduces the reflecting characteristics of the original scheme in order to ensure that light levels and brightness distributions are maintained at the original values.

These are the photometric requirements of the interior decoration scheme. There are also other attributes of colour which must be taken into account in a ward design. While the photometric requirements of the interior decoration limit to some extent the choice of colour which can be used in a ward, nevertheless by the intelligent use of the Munsell system of colour classification, a wide choice of colour is possible.

Colour is known to have psychological properties which add considerably to the pleasantness, cheerfulness, and intimacy of the ward design. During illness people are often confined to the ward and have no way of obtaining any form of visual relief. It is particularly important to plan the colour in a ward to give the greatest possible visual satisfaction.

In choosing the dominant colour in a ward, the effect which

such colour may have on the appearance of the room in general must be borne in mind. Excessive use of blue can so change the colour quality of the inter-reflected light in the room that the appearance of patients may be affected to some degree. The excessive use of red may be equally unsatisfactory.

The larger surfaces of the ward, the walls, ceiling, and floor, should be restricted in colour saturation. Light is inter-reflected between these surfaces in such a way that any excess of hue is exaggerated and so if they are all of the same basic hue, there can possibly be an unsatisfactory build up. It is preferable to restrain the colour on these surfaces and to supply any required colour stimulation by the use of a few judiciously placed areas of strong colour and also by the use of attractive bed covers and curtains which of themselves can create the necessary character of the room, but which can also be changed more easily to cope with any change in the nature of the ward, for example, from a physiological to a surgical ward.

The use of strong colour in a ward where the larger surfaces are restrained in colour will then have a more marked effect. Colour can be used to assist orientation. An ambulant patient may have difficulty in finding his way about in unfamiliar surroundings. The use of strong colour on the door attracts the attention and is therefore of assistance. It should not be necessary to point out that common sense should rule the use of such strong colour, but unfortunately examples have been seen where the same colour has been used on a door which the patients use and for another door which is intended for staff only.

Colour can be used more adventurously in the day spaces than in the sick wards. It is here that individual preferences and fashion often take over. Where fashion keeps in line with the requirements of good and comfortable vision, it is all to the good that it should change from time to time and there is no reason why the redecoration of a ward or a day space should not introduce a new colour scheme provided the photometric requirements are not affected. White paint around windows is good because it buffers the sky brightness and produces the right kind of contrast grading between the bright sky and the interior of the room. On the other hand, a fashion for black paint round windows must be condemned on visual grounds because it introduces harsh and uncomfortable contrast.

The British Standard Specification 2660 (colours for building and decorative paints) gives a selection of colours which have been chosen for their suitability for use in building interiors. The colours are divided into two groups, one of strong, bright colours to be used mainly to create centres of interest, and the other groups of relatively low saturation for use on large areas such as walls. The strong colours, if used on large areas, would so dominate a room as to make anything else subservient.

It is widely held that colour plays some therapeutic part and that it can therefore aid the recovery of patients in hospital. Although there is very little factual evidence for such therapeutic effects, this does not mean that they do not exist and that colour therapy should be discounted. The subject is an open one and no doubt will in due course receive proper investigation.

9

The Lighting of General Offices, Drawing Offices and Laboratories

The lighting of large general offices has improved considerably following the introduction of the fluorescent lamp. Efforts are being made continually to raise the standards of office lighting along with other factors which assist in the provision of comfortable and efficient working conditions. Property owners and developers recognise that the provision of good natural or artificial lighting is an amenity for which tenants are prepared to pay, and designers must now be prepared to provide lighting up to good modern levels using the best current practice.

ARTIFICIAL LIGHTING DURING DAYLIGHT HOURS

The improvements in office lighting have been more marked in artificial lighting than in daylighting, and there is an increasing tendency to use artificial lighting as the working illumination, confining the use of the window to providing a view and a visual rest centre. Because of the improvements in fluorescent lamp colour and efficiency, there is no longer an aversion to working in artificial lighting during daylight hours. Concurrent with the developments in lamp design are comparable improvements in lighting fittings which provide high levels of illumination on the work while at the same time illuminating the surroundings more efficiently and more pleasantly than was possible with the old type of concentrating fitting which left the room in semi-darkness.

The claim is often made, and is to an increasingly great extent justified, that high levels of working illumination in large offices are better provided by artificial means than by natural daylight. It is easier to control glare and directional characteristics of artificial lighting than it is to control natural daylighting in a large office where the same type of work has to be undertaken over a very large floor area. There are no advantages in top lighting by natural means as opposed to artificial means because exactly the same light distribution can be achieved artificially, and the difference in colour between overcast daylight and the best forms of fluorescent lighting is so small as to be imperceptible to all but those engaged on precise colour discrimination. Lateral daylighting still has advantages over artificial lighting by virtue of the modelling and sense of solidity which it gives, but good lateral daylighting can only be provided in the areas near to the windows and so the useful working area of the office is limited seriously in this way.

It does not follow that the future lies in total artificial lighting, even though in many other countries, particularly the United States of America, there is a trend in this direction. In Great Britain, however, and in north-west Europe generally, property owners can still place considerable monetary value on good daylighting when assessing the rental. This appears to be particularly true in densely built up city areas where good daylighting is difficult to obtain and is therefore prized as a valuable amenity.

The future in climates such as that of north-west Europe, as opposed to unfavourable and intemperate climates in other parts of the world, is to design offices so that the working light over the greater part of a deep room is provided artificially, while still providing good windows through which some daylight can penetrate and through which a view can be obtained. It must be borne in mind that even though the daylighting on desks remote from side windows may be minimal so far as the component on a horizontal plane is concerned, there will still be considerable daylighting on vertical planes which will modify the dominant downward component from the top lighting (see Chapter 6). The effects of this are much greater subjectively than would be predicted purely from the physical measurements of illumination level. Experiments undertaken at the Building Research Station have demonstrated that in a room in which the window area is no greater than one-sixteenth of the floor area, the natural lighting through this minimal fenestration nevertheless provides a component during daylight hours which is welcomed when it is present, and missed when it is absent. The room during daylight hours is a more pleasant and interesting place than the same room when daylight is no longer available.

Failure to appreciate that the subjective aspects of daylighting are at least as important as the quantitative aspects has led to misunderstanding between architects and artificial lighting interests. Windowless offices are not acceptable to the average office worker. In some circumstances there may be good reasons why the environment must be windowless, for example, in offices where maximum security is necessary; in such circumstances people are not so conscious of deprivation and are prepared to accept what would otherwise be considered unsatisfactory.

Judgements of the adequacy of office lighting that are based exclusively on sociometric surveys need careful interpretation. These surveys often show that no matter how poor the lighting, people in general express themselves fairly satisfied with what they have. Complaints of office lighting where it is not so woefully inadequate that people actually cannot see are usually based on discomfort from glare. These complaints are not usually expressed in this way but rather in the form of words such as 'too much light' or 'too bright'. The difficulty of interpreting this type of survey is that satisfaction changes with the prevailing environment. Twenty years ago a survey conducted by the Central Office of Information indicated that people were basically satisfied with the artificial lighting then provided mostly from filament lamps to a level of the order of only 5–10 lm/ft^2. In a more recent survey conducted by the Building Research Station, the same degree of satisfaction was expressed, though now the illumination levels were in the range of 10–30 lm/ft^2. A strict interpretation of these surveys would suggest that the level of artificial lighting is of no importance as far as subjective satisfaction is concerned. Yet other experience suggests that once people have been provided with higher levels, they are reluctant to go back to lighting which previously gave them satisfaction. Methods of appraising the subjective aspects of lighting are discussed in Chapter 13 (below).

STANDARDS OF OFFICE LIGHTING

The standards of illumination level on the work which are now recommended by the Illuminating Engineering Society range from 300 lux (30 lm/ft²) for general offices to 600 lux (60 lm/ft²) for drawing offices and business machine operation and 1000 lux (100 lm/ft²) for work which requires greater visual effort. These standards are met in very few old office buildings but in modern offices it is becoming more the custom to provide a general 'package deal' lighting installation to at least the level recommended by the I.E.S.

These recommended levels put forward by the Illuminating Engineering Society are based essentially on an experimental examination of the relation between illumination and the performance of tasks involving the recognition of critical detail and contrast. In practice these levels are probably approaching the optimum values for interior lighting with present methods of illumination from fittings mounted in the ceiling. The upper limit is not so much a factor governed by the level of lighting on the work as by the presence of large numbers of bright lighting fittings in the upper part of the field of view. The Illuminating Engineering Society couples its recommendations for working illumination with the recommendation that the Glare Index in offices should not exceed a value of 19. In order to achieve this degree of freedom from glare, lighting fittings must be designed to restrict the amount of light emitted sideways in the direction of the observer's eyes.

There are so far no statutory requirements in quantitative

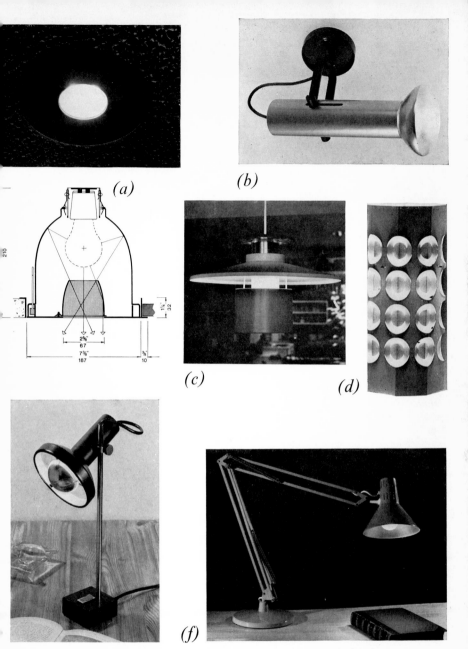

Plate 61. A wide range of fittings is now available for domestic use. (*a*) Pinhole fitting for recessing into ceiling. (*b*) The original Reid Universal floodlight. (*c*) Metal pendant fitting with blue painted drum, light grey 'ruff', and polished copper cap. The brightness of surfaces is graded by reflected light (Designer: David Medd). (*d*) Inexpensive cut-out paper shade for pendant or table mounting. (*e*) Spotlight for table or wall mounting using an ordinary 60 watt lamp. (*f*) An adjustable desk lamp in the 'Anglepoise' tradition. (Photographs: (a) Henk Snoek and (b) Christopher Moore by courtesy of Rotaflex Ltd.; (d) by Stables and Gray; (e) and (f) by British Lighting Industries Ltd. Studio Jaanus Ltd.)

Plate 62 (*a and b*).
The artificial
lighting of the
Schrieber House in
Hampstead helps to
underline its strong
formal qualities and
its lavish furnishing.
The symmetrical
arrangement of
windows emphasises
the 'palazzo' scale of
the building, both by
day and after dark.
(Architect: James
Gowan. Photograph
by John Donat.)

Plate 63 (*a and b*). The daylighting in this study bedroom at the Chichester Theological College continues the monastic tradition. The interest of the fixed skylight arises as much from the profile it gives to the building as from the flood of light it pours onto the desk, conducive to solitary concentration. There would need to be a blind on it to shield the student from the sun. The other window throws light to the back of the room and allows the student to commune with nature. (Architects: Ahrends, Burton and Koralek. Photograph by John Donat.)

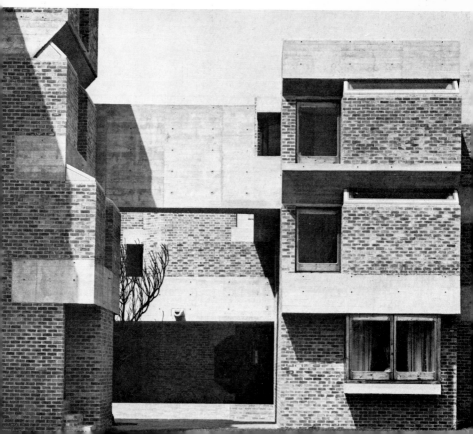

Plate 64. The dining-kitchen in this family house uses only local lighting of different kinds appropriate to each situation in the room. It can be used as an unobtrusive supplement to daylight during dull days. No diffusing shades are used. In the evening the luminances range from 25ft-L (270 asb) for the white plate on the table, to 3 ft-L (32 asb) for the table top, the pendant lamp, and the faces of people seated at the table, and 1–2 ft-L (11-22 asb) for the averages of the walls, floor and ceiling. (*a, below*) 1. 60 watt directional fitting focused on picture. 2. 100 watt pendant fitting in polished and painted metal above eye level, over dining table. 3. 3 × 60 watt tungsten strip lights behind pelmet over dresser. 4. 2 × 100 watt tungsten fittings recessed in dropped central ceiling separating kitchen and dining areas, and lighting work tops, glass storage shelves, and space where ironing is done. 5. 3 × 4 ft. 40 watt colour-matching fluorescent lamps behind pelmet over main kitchen working surfaces. 6. Covered porch facing south-west. (*b, above*). All walls are dark blue (B.S. 7–086) except for the white S.W. window and picture walls. The raised ceilings are light grey (B.S. 9–095), and the floor a light greyed yellow. (Architect: John Kay.)

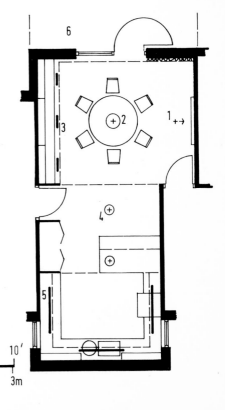

Plate 65. Outdoor playground lighting. The usefulness of playgrounds in residential areas can be extended by suitable floodlighting. The pitch shown is lit to about 2 lm/sq. ft. (20 lux) by 1000 watt lamps in open metal reflectors mounted on 25 ft. high posts. Similar playgrounds are now being lit with efficiency tungsten iodine lamps in more compact fittings. (By courtesy of the Greater London Council. Photograph by Mann Brothers.)

Plate 66. This sitting room forms an extension to an existing house. The window looks straight out into trees, and has vertical louvred blinds for sun control and use after dark. The artificial lighting includes a concealed strip above the textured matting on the walls, and pinhole crossbeam reflectors recessed into the boarded ceiling. (Architect: John Reid. Photograph by Sam Lambert.)

CHOICE OF LIGHT SOURCES

Fluorescent lighting is indicated for use in offices everywhere except where in individual offices personal preference may be expressed for filament lighting. In small offices which the workers maintain themselves, also, the advantages of simplicity and easy maintenance afforded by filament lighting are obvious. In all other places well-designed fluorescent lighting is so much more satisfactory on all counts that its choice is always indicated.

The choice of type of lamp must be based entirely on requirements for colour rendering and upon personal preferences. Where the nature of the office work, for example in a drawing office, requires critical judgement of colour, the type of lamp known as 'colour matching' or 'artificial daylight' may be indicated. This lamp has a somewhat lower luminous efficiency than the most efficient type of fluorescent lamp and consequently for economic reasons it will not be used unless there is a clear requirement.

The choice will then lie between types of lamps which are basically similar to daylight in colour and lamps which have been deliberately designed to give a warm colour to match that of filament lighting. During daylight hours such 'warm' types of lamps appear pinkish or yellowish. Where the installation is to be used more frequently during daylight hours than at night, it is preferable to use either a 'daylight' type of lamp, a 'natural' type or a 'white' type, all of which are similar to daylight in different degrees. Where staff are particularly conscious of colour changes which affect the appearance of the complexion, etc. lamps with a better colour rendering (often branded 'de luxe') may be indicated. As lamps improve and as people become more familiar with the kind of colour rendering given by fluorescent lighting, fewer complaints are received on the score of the colour of the light.

The choice of light source should also be related to the choice of decoration and colour in the office. Surface colours of high reflectance are necessary both to enable the maximum use to be made of reflected light in the room and also to ensure a sufficient background brightness to relate satisfactorily to the brightness of the work. Some office staff are more sensitive to the colour treatment than to the lighting. There is at the moment a tendency to use austere and restrained colour treatment in offices which, together with the type of colour rendering given by much fluorescent lighting, has an institutional and unfriendly 'cool' appearance. In a survey undertaken by the Building Research Station, it was found that colour treatment in offices was often rated as 'too cold'. Any choice of decoration scheme for an office must be planned in relation to the type of fluorescent lighting to be installed. A colour scheme which is attractive and warm looking when illuminated by filament lighting, or by daylighting, may lose its attraction under 'white' fluorescent lighting. Some guidance exists for the choice of colour in relation to lighting but individual designs often have to be treated on their merits with expert assistance (see Chapter 5).

PERMANENT SUPPLEMENTARY ARTIFICIAL LIGHTING IN OFFICES

Because of the need to make the fullest use of site space in congested urban areas, offices must be designed to make the fullest use of P.S.A.L.I. The principles of P.S.A.L.I. require that there should be sufficient daylight admitted for the windows to create the lighting character in the room, the artificial supplement being an additional component of working illumination particularly on the desks remote from the windows.

In practice the concept of P.S.A.L.I. is in rivalry with the concept of deep offices with predominantly artificial lighting rather than predominantly natural lighting. It is probable that staff will become more used to working in predominantly artificial lighting situations, although this would not be their choice if they had one. There is no doubt, however, that in deep offices with artificial lighting, the presence of a window, even though this may provide only a small component of daylighting, is welcomed.

The only way in which good quality lighting can be achieved to the levels now recommended is by a proper integration of daylight and artificial light, or otherwise by artificial light alone. In the integration of daylight and artificial light, the daylight must be taken as supplying the character or amenity of the lighting and the artificial light much of the working illumination. The levels of supplementary light must be determined by a consideration of the adaptation of the eyes when working in the proposed environment. In practice artificial lighting in the range 500–700 lux (50–70 lm/ft²) is found to be satisfactory as a supplement to office daylighting, so that in fact the levels which are satisfactory from the point of view of balance of brightness between the day-lit areas and the artificially supplemented areas are slightly higher than the order of illumination required by the Illuminating Engineering Society recommendations. Preferably the artificial supplement should be integrated into the building structure in such a way that the worker is unconcerned that he is working in partial artificial light. The designer must also take into account the likelihood of the room layouts and partitions being altered from time to time.

DRAWING OFFICES

It is said that of all 'white-collar' workers, draughtsmen have the highest incidence of spectacles, and that this is an indication of the visual difficulty of their job. It may also be indicative of the unsatisfactory conditions under which too many of them work. Poor daylighting is often supplemented by makeshift artificial lighting.

The draughtsman's first requirement is for an adequate level of well distributed light over the whole of his working plane which may be a fully adjustable tilting drafting machine. His light should be free from glare, and supplemented by preferential lighting on his work, if possible under his own control. It would seem that at the moment the lighting of drawing offices divides draughtsmen into two camps, those who demand adjustable lights by their boards, and those who prefer to rely on general lighting alone.

Sociometric surveys in drawing offices, particularly work undertaken by the Building Research Station, have not given a positive answer to decide which of these two philosophies is correct. On the other hand, direct analytical experimental work on visual acuity and visual comfort has demonstrated that vision is best, that is, more efficient and more comfortable, when the work is slightly brighter than the immediate surroundings which in turn should be slightly brighter than the general surroundings. This would argue strongly in favour of those who prefer to have their own adjustable local light, provided that this does no more than supplement good general lighting in the whole of the drawing office.

THE LOCAL LIGHT

When the results of both approaches are taken together, however, and the data of the sociometric studies are examined in relation to the results of the experimental study, it would be true to say that the two are not inconsistent. Where draughtsmen are provided with inadequate general lighting, they then insist on having a personal local light. On the other hand, where draughtsmen are provided with good general lighting, they are then happy to rely on this lighting alone. None of the offices in the social survey were provided with the combination of good general and local lighting prescribed by the experimental investigation. If they had such lighting, they might well prefer it to anything which they had before.

While the local light is the traditional solution, probably because it has usually been the only means by which the draughtsman could get sufficient illumination on his work, it has drawbacks particularly in mechanical inconvenience. Unless great care is taken in the design of the local light, it can not only be a source of irritation to the draughtsman himself, but it can be left in positions which will dazzle others on nearby boards.

Another limitation of the local light is that it is not likely to be conveniently placed for the other work which a draughtsman has to do away from the board. The Building Research Station survey showed that only one-third of a draughtsman's time is actually spent on drawing, and that a considerable proportion of his time is taken up with calculations and with reference to other drawings. These are operations which also require good lighting for speed and accuracy and therefore argue in favour of adequate general lighting in the office for all but the most exacting visual tasks at the board.

The conclusion from both survey work and the experimental work conducted at the Building Research Station is that drawing office lighting is best provided by a high level of good overall lighting of not less than 30 lm/ft² (300 lux) and preferably up to 50 lm/ft² (500 lux). In addition to this good general lighting there should be available, should the draughtsman require it, local lighting to give him up to 150 lm/ft² (1500 lux) locally on his board. An important principle is that the areas where the most exacting work is done should appear brighter than anything else within the draughtsman's visual field because this accentuates the work to attract his attention and to enable him to concentrate with less strain amid the distractions often present in a large office.

The draughtsman himself can often alleviate the difficult visual conditions of his work by providing partial cover to his board with paper or material of reflectance of the order of 30%. Large areas of white paper illuminated to high levels of illumination are themselves a source of direct glare. So long as draughting has to be done on white paper, the only way the visual conditions can be modified is to reduce the area of white in the way suggested. Attention to the choice of reflectances of other surfaces in the office, including the floor and furniture, will also assist in ensuring that there is the proper gradation of brightness from the work itself, which should be the brightest, to the general surroundings, which should be the least bright part of the visual field.

DAYLIGHTING

Daylight appears to be the preferred illumination for draughtsmen, but this may be because they have never had the opportunity of working in really well designed artificial lighting. To give satisfaction, where daylight is intended to be the main working illumination, it must be evenly distributed throughout the whole office, otherwise staff working at drawing boards some distance from the windows will feel deprived of light. Draughtsmen feel that the distribution of daylight is as important as its quantity and they prefer that the daylight should come from a wide angle. Daylight coming from small windows is highly directional and throws hard shadows. If such directional light comes from a bad angle in relation to the work, it can be a cause of visual strain and can lead to error.

A Daylight Factor of 5% is recommended as the minimum level for drawing offices, and it is a value which appears to enable the majority of draughtsmen to work at their boards without the need for artificial light except on dull days. Daylight of this level cannot be easily achieved by side lighting, particularly in urban areas. Very large side windows cause high contrast and glare. The daylight can best be provided by roof lighting well recessed and louvered to prevent direct sunlight coming through where it is not wanted. All windows, whether in the walls or in the roof, should be provided with blinds or curtains so that the daylight can be controlled both in quantity and in direction. If the working daylight is provided from roof lights, nevertheless there should be some side windows to provide a view outside because people who are engaged in work of great visual difficulty require visual rest centres, that is, a view on which the eyes can relax their accommodation and convergence for a brief period before resuming muscular effort.

ARTIFICIAL LIGHTING

Artificial lighting can be designed to perform exactly the same function as daylighting, particularly if it is well integrated with the building structure. It is possible to design a system of fluorescent lighting recessed in the roof which is indistinguishable from natural lighting, or which can be integrated with a natural lighting system in such a way that during daylight hours the daylighting can be the dominant working light, to be supplemented and then taken over by the artificial lighting as daylight fades.

Such a system of artificial lighting should provide an average level of illumination not less than 500 lux (50 lm/ft²). During the greater part of the working day, especially during spring, summer, and autumn, a Daylight Factor of 5% will provide not less than 500 lux (50 lm/ft²) and so the draughtsman will become accustomed to working in such levels of illumination. If anything less is provided by the artificial lighting, there will be a sense of deprivation and inadequacy.

FLUORESCENT LIGHTING

Fluorescent lighting is indicated for the lighting of drawing offices. The Building Research Station sociometric survey demonstrated that where fluorescent lighting is installed, more than 70% of the draughtsmen interviewed were able to register approval with or without further qualifications. The most common objections to fluorescent lighting concern glare. Modern lighting techniques can avoid direct glare and it is likely that those who complained were suffering from badly designed lighting.

Flicker. If fluorescent lighting is used in drawing offices, it is important that the engineering must be skilful and in particular there should be no risk of flicker. The eye is progressively more sensitive to flicker the higher the brightness and so under the well lit conditions in a drawing office, the fluctuation in intensity of the lighting with the alternating current cycle will be perceived as flicker, unless special lighting circuits are employed. Preferably lamps should be wired on the three phases of the A.C. distribution system and suitably interlaced in their spacing, but twin lamps on a standard 'split-phase' (lead-lag) circuit are often quite satisfactory with most types of lamp.

The type of fluorescent lamp to be chosen will depend upon this problem of flicker to some extent. If an illumination of the order of 500 lux (50 lm/ft²) is to be provided, it is likely that a lamp chosen for its suitability for matching daylight will also be acceptable at night so far as colour appearance and colour rendering are concerned, but unfortunately some recent improvements in lamp colour rendering properties have been accompanied by a greater tendency to flicker. If it is not possible to employ split-phase or 3-phase circuitry, it is then more important to select a type of fluorescent lamp which employs a luminescent powder with a long 'after-glow'. Such a selection can only be made in consultation with the lamp manufacturer at the time because the types of luminescent powder are constantly changing and vary from manufacturer to manufacturer. Colour appearance is no clue.

'End-flicker', the 50-cycle alternation around the electrodes of the tubes, can be eliminated by the use of screens over the electrodes. This is standard practice in good light fitting design.

Glare. Glare from artificial lighting and from daylighting also must be kept down to a minimum. The Illuminating Engineering Society recommends a limiting Glare Index of not greater than 19 in drawing offices. Greater visual comfort will result if this can be reduced to 16. A Glare Index of 16 can easily be achieved by recessed louvered lighting, although it is also possible to meet the requirement from suspended lighting of a design which allows enough light on the ceiling and limits the brightness in the direction of view. The higher the illumination provided in an office, the more critical will be the balance of brightness. No lighting fitting in the field of view should be excessively bright. The luminance seen at any angle less than 45° above the horizontal should not exceed 5000 asb (500 ft-L) and should preferably be much lower. If the lighting is in the form of an overall luminous ceiling or of individual louvered areas inset in the roof, the maximum luminance should not exceed 1500 asb (150 ft-L).

Much better results are achieved if the brightness of the ceiling can be kept down to a value not greater than one half that of the white paper on the drawing board. This is not difficult to achieve if louvers or baffles with a reflectance of the order of 30–50 % are used.

While direct glare from the lighting is not difficult to eliminate with skilled design, it is more of a problem to avoid unwanted reflections on the board from ceiling mounted lighting. The Standards Association of Australia (Australian Standard CA 30/1957) recommends that the illumination on the drawing board should be related numerically to the brightness (luminance) of those parts of the lighting units reflected in the surface of the board. (This Australian recommendation is that the illumination on the work in lm/ft² should be not less than twelve times this luminance in candelas/in² of the lamps, i.e. the illumination in lux should be not less than one-fortieth of the luminance of the reflected image in apostalbs.) In practice this Australian recommendation if interpreted strictly may call for an unnecessarily high general illumination of the whole office. In practice it is more satisfactory to rely upon local preferential lighting which can bring up the illumination on the work itself to a value of up to 1000 lux (100 lm/ft²). If this is done, use can be made of the directional characteristics of the local lighting and it is then easy to swamp any unwanted reflections caused by the general lighting. Reflected images can sometimes be alleviated by careful placing of the draughting boards in relating to the positions of the lighting in the ceiling.

Tracing is an activity of particular visual difficulty which is best assisted by special lighting. Adjustable local lighting is particularly valuable here, both to swamp reflections from the general lighting in the glossy tracing linen, and also to permit the tracer to set the light to give the best visibility for his or her own particular requirements. Transilluminated tracing tables can sometimes be used but are not always preferred because such a large area of relatively high luminance can cause visual discomfort. If a transilluminated tracing table is indicated because of the nature of the work, visual discomfort can be reduced by covering up those parts of the table which are not in use at any given time, bearing in mind that glare discomfort is not only a function of the brightness of the glaring source but also of its area. Reducing the glaring area of a tracing table to the area in use at any given time, say to one-tenth of its total area, can make a very substantial reduction in the degree of glare discomfort.

OFFICE LANDSCAPING

Brief reference was made earlier (Chapter 6) to the office design technique know as 'bürolandschaft' or office landscaping, which was developed in Western Germany. The whole of a large space, perhaps as much as 70 metres (200 ft.) square or even greater, is used without any permanent partitions by the whole of an office from the most junior up to all but the most senior staff. The essence of the technique is to deaden the noise of office machinery by use of thick pile carpet, acoustic treatment of walls and ceilings, and by the use of acoustic screens around areas which require privacy or segregation. Indoor plants and other decorative features supply the 'landscaping' and view windows around the periphery of the space give some visual access to the outside world.

The lighting in such offices is invariably by fluorescent lamps inset into the ceiling or otherwise integrated with the structure. When well done the effect is clean, hygienic, and Central European in the best sense of the term, on all counts is far superior to the dull, jumbo-sized offices with monotonous rows of fluorescent trunking seen all too frequently elsewhere.

Office landscaping would well repay careful study by lighting designers, because it has many advantages which could be developed with the necessary understanding of basic lighting principles. It is true that uniform lighting throughout the space permits complete freedom for change, which is one of the advantages claimed for the technique, but in practice it is unlikely that after the first few months of change-about there will be many upheavals very often. Consequently further advantage could be gained by the use of light to create the special character of certain areas, such as those used by staff who have to conduct interviews or by senior executives, which at the moment are marked out only by the acoustic screens and the use of plants. Tasks of special visual difficulty could also be given preferential lighting simply to hold the attention in distracting surroundings, even if the level of general lighting was fully adequate.

THE LIGHTING OF LABORATORIES

With the growth of industries based upon technology, more buildings are being designed as laboratories, to be used for a wide variety of activities. Some of these activities are so little different from those performed in schools or offices that they demand no special discussion, while others are development-factories demanding highly individual design.

Many laboratories undertake work which requires accurate colour discrimination. In such rooms natural daylighting is indicated where possible because the colour rendering given is familiar and the eye can adapt easily to the slight variations in the spectral composition which occur throughout the day. Work involving accurate colour judgements should be reserved for occasions when daylighting is adequate. The best form of fluorescent lighting for the purpose, the 'colour matching' or 'artificial daylight' type of lamp, is an excellent substitute for natural daylight, but even so the colour rendering is not precisely the same and requires a certain amount of experience before colour judgements can be made precisely the same as those made under natural daylighting. For many laboratory tasks, however, it is perfectly adequate.

Much laboratory work also involves a high degree of visual precision (Fig. 20). Very high levels of illumination which may be indicated cannot easily be supplied by general lighting throughout the laboratory. Such work is better zoned to take place near windows, or alternatively special local lighting should be provided. It is both uneconomical and unsatisfactory visually

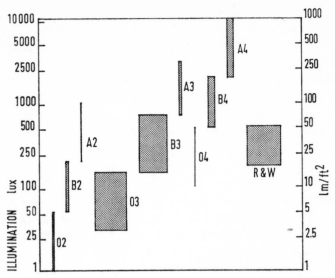

Fig. 20. Illumination values for laboratory tasks. The diagram is based, with modifications, on Fig. 69 of 'The Design of Research Laboratories' (Nuffield D.A.S., Oxford 1961). The width of the blocks shows the proportion of laboratory time spent on tasks of different contrast (2—good, 4—poor contrast) and visual difficulty (0—easy, A—minute detail). Thus most of the time is spent on R and W (reading and writing) or on tasks of little or average difficulty and contrast (0—3 and B—3). Consequently, a moderate level of general lighting can be provided; difficult work is done at the window or by local lighting.

to provide such very high levels of illumination over the whole laboratory area, and in any case visual performance of the difficult tasks will not be as good as when local lighting is supplied on the work alone.

In some special circumstances, particularly where close control over temperature and humidity of the environment is essential, it will be necessary to provide a totally artificially lit environment. Experience with windowless laboratories, though not extensive, indicates that levels of illumination should be high, not less than 500 lux (50 lm/ft²), if a feeling of deprivation is not to result when people go into the laboratory from outside to begin their working day. Once in the room, the eye will adapt to the prevailing illumination and will be unaware of outdoor conditions, provided there is no opportunity to see out or go out during the working period. Nevertheless a high level of illumination is indicated because such evidence as can be obtained about working in windowless enclosures (particularly from Scandinavia, Russia, and the United States of America) indicates that people suffer less physiologically and psychologically if the interior windowless environment is provided with a high level of lighting, distributed to give an impression of brightness.

Fluorescent lighting can be used in laboratories but care must be taken to ensure that the right type of lamp is used consistent with the requirements of colour rendering. Where the visual tasks do not demand the 'artificial daylight' type of lamp, the more efficient 'daylight', 'natural', or 'white' types of lamp can be used with advantage.

P.S.A.L.I. in laboratories follows the same principles as in offices or factories, except that special attention may have to be paid to the directional characteristics of the lighting in certain circumstances. For example, titrations are often better undertaken against the background of a well illuminated white wall rather than by direct lighting on the equipment itself. The lighting of a row of instrument dials must be so arranged that disability glare is not caused by unwanted reflections in the instrument cases. The manipulation of equipment in cavities requires controllable local lighting together with good general lighting. The detection of flaws or interfaces in an otherwise plane surface may be assisted by a thin, flat beam of light adjusted to the surface at glancing incidence.

It has to be borne in mind that in laboratories, as in many other types of building, the lighting must be related to the building design as a whole. In the case of laboratories the shape of the room is often governed by the way in which the special supply services and the benches are arranged. Work undertaken by the Nuffield Foundation Division of Architectural Studies revealed that in many laboratories there are advantages in rooms of some considerable depth running in from the window wall. Such rooms cannot be lit economically by natural lighting but require instead good natural lighting supplemented by artificial lighting at the back of the room. In such deep rooms the level of general lighting can be set to cater for the bulk of the moderately difficult visual tasks, selective local lighting being provided for tasks which demand special attention.

Many laboratories demand special and individual design of the building as a whole, taking into consideration the unique problems raised by the special nature of the work. The lighting is only part of this design, but it is essential that it should be integrated with the whole at an early stage, because at a late stage it may be impossible to provide exactly the right lighting for the job. This is especially the case where the lighting needs to be built into the structure, as may be the case where the utmost cleanliness and freedom from dust and bacteria are essential.

10

The Lighting of Factories

Direct returns, in terms of productivity, for the installation of good lighting are to be found in factory buildings. The direct influence of good lighting on productivity has been clearly demonstrated and so the present standards of the best factory lighting probably owe at least as much to this as they do to the desire of management to provide pleasant surroundings and promote industrial health.

When factory lighting was universally of a poor standard, it was very easy to demonstrate that improvements in the lighting would effect a considerable reduction in accidents, and an improvement in both the quantity and the quality of the product. The law of diminishing returns, however, operates with the result that, as lighting has continued to improve, the direct effect on productivity becomes less and less directly measurable. The claims of good lighting in factories begin to rest more upon amenity. Nevertheless, management all over the world is convinced that, in the long run, the lighting pays for itself in many ways, not forgetting greater work satisfaction, less absence due to minor sickness and reduced labour turnover.

In considering the amenity aspects of industrial lighting, the decorative treatment and subsequent maintenance of the building itself must also be taken into account. Clean, light surfaces not only play their part in reflecting daylight and artificial illumination, and so rendering the whole installation more efficient, but they also create a lively and pleasing environment, where the judicious use of colour plays an important part.

Factory lighting is controlled by the Factories Acts and Regulations. There is a Statutory Instrument (Section 5 of the Factories Act, 1937) which demands effective provision for securing and maintaining sufficient and suitable lighting in factories. The Ministry of Labour has laid down certain standards, which are incorporated in the Factories (Standards of Lighting) Regulations, 1941, which specify minimum standards of lighting level, and also give indications on precautions to be taken for the minimizing of glare. These Regulations fall a long way short of good current lighting practice, but nevertheless their existence in this form enables the Ministry of Labour to have factory premises inspected to ensure that both the natural and artificial lighting is not seriously deficient.

The Factories Act, 1937 permits these Regulations to be revised from time to time, in the light of current practice and new industrial processes and, if necessary, on the grounds of general health and amenity.

PRINCIPLES OF FACTORY LIGHTING

It is very desirable to consider the lighting of a factory area in two stages. First, there is the general *building lighting*, which is provided to enable people to go about the building safely and in comfort, and also to provide a standard of amenity appropriate to the environment. Second, there is the *work lighting*, which may be special local lighting on machines and work areas, or which can, in some cases, be integrated with the building lighting.

The building lighting can be provided by natural daylight or by artificial sources, or a combination of both.

The work lighting will in general be provided by artificial sources, because of the ease of control and direction of light. Some processes, however, must be undertaken in natural light, and where this is the case the local work lighting will be provided by windows of special design, or by zoning the work in such a way that daylight of the required amount and directional characteristics is provided at the work.

16

QUANTITY OF ILLUMINATION

The amount of illumination which is provided in a factory will therefore be linked to the building and to the work. The Illuminating Engineering Society has recommended, in its current Code of Recommended Lighting Practice, that all factory areas should receive a minimum illumination of not less than 200 lux (20 lm/ft²) on the grounds of amenity alone. This recommendation is in line with current practice in well lit industrial areas, even though the Factories Act calls for a minimum illumination of only 6 lm/ft² (60 lux).

The I.E.S. Code also recommends that the level of daylight in an industrial area should not fall below that represented by a Daylight Factor of 5%. This provides the amenity level of 200 lux under a sky of 4000 lux, which is a rather dull sky. On brighter days the level will, of course, be higher. The artificial lighting 'amenity level' will, therefore, be less than the basic daylight level except on dull days.

DETERMINATION OF THE WORK LIGHTING BY
ANALYTICAL METHODS

Vision is improved up to a point by providing more light on the work. In Chapter 3 the relation was explained between lighting levels and the two critical factors (a) the size of detail and (b) the contrast which reveals detail. Every task which is undertaken in a factory can be broken down into its critical detail and critical contrast. Thence, it is possible to prescribe the amount of illumination necessary on the task for an agreed standard of performance.

The *standard of performance* is more difficult to define. The Illuminating Engineering Society, in its Lighting Code, has followed the work of Weston and others in which visual performance is defined as a fraction of the performance possible under ideal lighting conditions. For example, a given factory task may be performed without error in a given time, in ideal lighting. The limitation is then on manual and other factors, and not visual factors. If the illumination on the task is less than the critical value for maximum performance, the performance will be less than the best and the reduction will be due to the inadequate lighting. In this way, by carefully designed experiments, it is possible to relate the light on a task of given critical detail and contrast, with the percentage visual performance expressed in terms of the time taken for best performance to the time actually taken with the lower level of lighting. Weston's definition of visual performance is, in fact, rather more complex than this, because he had to take into account not only the speed of performance but also the errors. However, this description will serve to indicate the method of assessing visual performance by Weston's technique. From this, it is possible to analyse a given factory task in terms of critical size and contrast, and thence to define the level of illumination necessary to achieve a given visual performance of that task. Until recently, the Illuminating Engineering Society aimed at 90% of optimum performance. More recently, a higher standard has been set. However, when levels of illumination beyond the 90% performance limit are considered, the Weston method of prescribing illumination level tends to become unreliable, but the method is sound and the ideas behind it still serve, in large measure, to define the amount of light necessary on the work as specified in the I.E.S. Code. This Code lists a very wide range of factory tasks and the corresponding level of lighting on the work.

GLARE AND ITS ELIMINATION

Glare is the bane of factory lighting, whether by daylight or by artificial lighting sources. The basic design of factory lighting equipment has been geared for so long to rigid economy that a tradition has grown up for austerity in factory lighting design not unrelated to the 'muck and brass' philosophy. Visual tasks in factories are often as exacting as anywhere else, yet whereas in an office the elimination of glare is accepted as part of the environmental design, in a factory is it often looked upon as an expensive trimming.

Both *disability glare* and *discomfort glare* can occur in a factory. Disability glare occurs when a bright light source is in close proximity to a delicate visual task. It is caused by the scattering of light in the eye, and by the reduction in sensitivity of the retina which results from the presence of the bright light source. Many examples occur in factory practice, but perhaps reflections in a polished piece of machinery are the most common cause in an otherwise well-lit factory. Disability glare is caused chiefly by inadequate quantity of light on the work coming from lighting fittings which send too much light in the eyes. Discomfort glare, on the other hand, often arises from the provision of too much light.

The Illuminating Engineering Society's Code of Lighting Practice gives limiting values of Glare Index for a wide range of different factory environments. These values of Glare Index are not linked closely to the nature of the visual *task*, but to the *environment* in which the task is likely to be undertaken. The Glare Index therefore applies to the *building lighting*.

THE DAYLIGHTING OF FACTORIES

The first requirement of the daylighting of a factory is to provide the building lighting during the daytime, either in whole or in conjunction with a permanent supplement of artificial lighting. The daylight can be determined by the illumination requirements of the visual task or it can be designed entirely from the point of view of amenity and visual comfort.

Design from the point of view of the precise requirement of the work militates against flexibility in the use of the factory. In practice, it is preferable to obtain a broad idea of the different types of work which will be undertaken in the building, from which a general idea of the necessary levels of working illumination can be found. If the work is all of one kind, and requires only a moderate level of working illumination, the daylighting can be geared to this visual task. On the other hand, if the work is varied, some requiring a great deal of light, it might be unwise to design the whole factory for the requirements of these most difficult tasks. Either the work can be zoned in such a way that the difficult tasks are undertaken in the areas of greatest daylight, or more satisfactorily, additional local lighting can be provided for those tasks which require more than the average illumination of the building lighting.

It is quite practicable to provide a uniform level of daylight over a roof-lit factory building of the order of 5% Daylight Factor. Little difficulty is experienced in increasing this level to 10%. An average level much above 10% involves large areas of glazing which will inevitably give rise to troubles due to excessive solar heat gain and discomfort from direct sunlight, during the summer, and probably to cold down-draughts and loss of internal heat during the winter. Since the minimum amenity level recommended for factories is 5% Daylight Factor, it is clear that there is very little flexibility in the choice of level of daylight, and it will be generally found that a *maintained* value of between 5 and 10% covers the practicable range.

It is difficult, or impossible, to obtain levels of average Daylight Factor of this order with other than lighting inserted in the roof. Well-designed lighting from two or more vertical sides can indeed achieve an average level of 5% or more, but the level will vary greatly over the interior, from 20% or more near the window to 3% or less in the more remote areas. Side lighting has its place in industrial areas, but the great majority of factories rely on roof lighting.

Roof lighting is only possible in single-storey factories; in other areas the preferable solution is side lighting combined with a matched supplement of permanent artificial lighting.

FORMS OF ROOF LIGHTING

The distribution of the building lighting should be as uniform as practicable in order to obtain the greatest adaptability of the floor space. Various forms of roof lighting are in common use

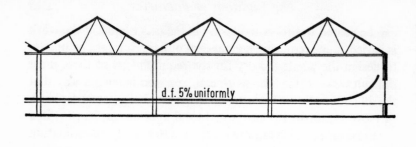

d.f. 5% uniformly

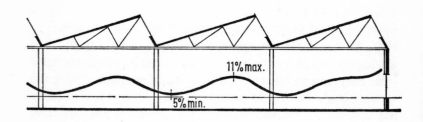

11% max.

5% min.

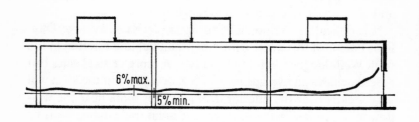

6% max.

5% min.

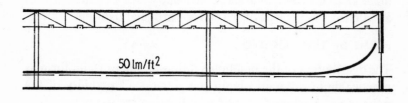

50 lm/ft²

for industrial areas. Of these, the shed-type roof, the north-light roof, and the monitor roof are employed most frequently (Fig. 21). Horizontal glazing and monitors with sloping glazing are also employed occasionally.

The *uniformity of illumination* is a function of the closeness of the spacing of the glazing apertures.

The *amount of illumination* is dependent upon the proportion of glazed area, and to a lesser extent upon the arrangement of the glazing in the roof.

EFFICIENCY OF DIFFERENT TYPES OF ROOF LIGHTING

The different forms of roof lighting can be compared on the basis of the amount of light which they provide on the working plane in the building. There are also other bases of comparison. The north-light roof, which give unilateral lighting in a direction away from the sun, has the advantage that it protects the workers from discomfort due to direct sunlight at all times of the year. The monitor roof with vertical glazing can be more easily kept clean, because the vertical glazing is less subject to the deposition of industrial dirt, the exterior is subject to the washing effect of rain, and both internal and external surfaces can be cleaned from catwalks more easily than other forms of roof lighting.

These secondary effects apart (although they may be primary under certain conditions) the most efficient form of glazing in terms of utilisable light flux on the working plane for a given

---

Fig. 21. This diagram of the main factory roof types is based on information given in the B.R.S. Factory Building Studies No. 2 (see Bibliography, p. 312). The B.R.S. Publication points out that if there are very dirty atmospheric conditions or if an extensive installation of overhead services is expected, then larger glass areas will be required to maintain the same illumination levels.

(a) Shed type roof with continuous strips of glazing. Glass area 10% of floor area
(b) North light roof giving unilateral lighting. Glass area 20% of floor area
(c) Monitor roof with vertical glazing. Glass area 30% of floor area
(d) Artificially lit factory with view window

amount of glass in the roof, is horizontal roof lighting. A shed-type roof is almost equally as efficient. Vertical monitor roof lighting will, for the same area of glazing, give only about one-half to one-third of the interior illumination given by shed or horizontal roof lighting. North lighting and sloping monitor lighting lie between these two, north-lighting giving about three-quarters of the light given by the same area of glass with continuous horizontal roof lighting. (See Table on p. 106.)

On the other hand, it must not be forgotten that horizontal and low-pitched shed glazing will admit sunlight and so cause discomfort unless some form of anti-sun screening is introduced below the glass. Once this screening is introduced, the efficiency of the system is reduced considerably, resulting in a lesser advantage for these forms of glazing. When making a decision as to which form of glazing is to be used, amenity and freedom from discomfort, as well as quantity of illumination, must be taken into account.

DISTRIBUTION OF ILLUMINATION

Different forms of glazing give different degrees of variation in the level of the light on the working plane. North-lighting will give a non-uniform distribution, a diversity of 2:1 being typical. Shed-type and horizontal glazing give little variation, provided the pitch of the lighting, i.e. the spacing of the rows of glass, is sufficiently close. If the pitch of the light is not greater than the height of the lights above the reference plane, reasonable uniformity will be achieved with horizontal or shed-type lighting.

Uniformity of illumination can be greatly assisted by the provision of internal surfaces, walls, floor, etc. of high reflection factor. The importance of interior decorations is at last being recognised by factory management, and it is no longer necessary to labour the point about the value of clean buildings. It is not always realised, however, that to provide a floor of high reflection factor is one excellent way of greatly assisting the uniformity of light distribution. Floor coverings of reflection factors of the order of 40% are easily obtainable and can be maintained without undue difficulty in all but the most dirty industrial areas.

SPECIAL FORMS OF GLAZING

Some factories have been provided with what is called 'distributed lighting' or 'starlighting', which consists of a very large number of small areas of glazing distributed over the factory roof. This system gives very uniform illumination but at the cost of a disturbing pattern of light which tends to act as a visual distraction. When the glazing is well inset in the roof, however, a louvering effect results which largely overcomes the objection. This method of glazing can then be evaluated in relation to other methods, entirely on the score of cost.

Corrugated resin-bonded glass fibre roofing materials offer an inexpensive method of glazing for distributed lighting particularly in temporary buildings. The material, which consists of corrugated sheets of partially diffusing plastic, can be integrated with a corrugated asbestos roof with the minimum of difficulty. Some varieties of the material are very subject to weathering effects, with the result that both the transmitting and diffusing properties, as well as the colour of the light admitted, suffer considerable changes. The material has the advantage that it diffuses sunlight, and so goes some way to reduce discomfort due to direct solar heat. The level of interior illumination produced by roof lighting glazed with material of this type can be found very simply. The procedure is to evaluate the Daylight Factor on the assumption that the opening is unglazed, and then to multiply this Daylight Factor by the total diffuse transmission factor of the material. This transmission factor can be obtained from the manufacturer. It is also recommended to allow for the effect of depreciation in use, because some of these materials are subject to weathering.

SUNLIGHT

Special care has to be taken to eliminate unwanted sunlight in a factory area. Most factory work is done from fixed positions, from which staff cannot move during the course of their work. Direct sunlight on the worker causes irritation and discomfort, and can thereby lead to accidents and reduced productivity. On the other hand, total elimination of sunlight from all areas of the factory is generally not liked. The windows should be so designed that some sunlight is admitted but only on areas where it does not interfere with work.

The customary north-lighting (or south-lighting in the southern hemisphere) eliminates direct sunlight during the whole of the year except in the early mornings and late evenings of the summer months. It is, however, during the winter, when sunlight is at a premium, that staff welcome it most. An alternative solution to the factory lighting problem was put forward some years ago, and is known as the B.R.S. Monitor roof. This is a system of monitor lighting with the main north face at 60° slope, and with a smaller vertical south face, designed to give a little sun penetration, but not sufficient to cause excessive solar heat gain during summer. The distribution of light from an overcast sky is fairly satisfactory, particularly if the roof and details of the monitor are painted white, and provided proper attention is paid to reflected light from floor and wall surfaces. The amount of the south-facing glazing can, of course, be modified to meet particular sunlight demands for different latitudes.

EFFECT OF INTERNAL OBSTRUCTIONS

The design of factory roof lighting depends upon the light being allowed to penetrate to the reference plane freely without obstruction. Clearly if there are obstructions at high level, due to ventilating ducts, girders, crane gantries, and, indeed, large artificial lighting systems, the actual amount of illumination which will reach the reference plane will be reduced. Some of these internal obstructions are inevitable, and cannot be dispensed with, but much can be achieved to restore the daylight by painting them with light colours, and maintaining the high reflection factors by constant cleaning and repainting where necessary.

ZONING OF WORK IN SIDE LIGHTING

In general, side lighting has more disadvantages for industrial buildings than roof lighting, but it is of course the only form of lighting in multi-storey buildings. It is very difficult to achieve uniformity of illumination over large areas of room lit from vertical windows in side walls. The utmost reliance has to be placed on the use of reflected light from floor, ceiling, and walls to provide a reflected supplement to the direct skylight.

The problems of side-lit factories are the same, in essentials, as those of side lighting in every other building. In a factory, however, where work of different visual difficulty may be undertaken, it may be useful to consider the zoning of the work in terms of visual difficulty. The different visual tasks can be analysed in terms of critical size and detail, and those which make the greatest visual demands can be placed near the windows and those which are visually less exacting can be placed in the less well-lit areas. This is a technique which has been followed in the design of laboratories (Fig. 20, p. 237) but in a factory where it is more usual for tasks of the same kind to be undertaken all in the same room, it may be more profitable to consider permanent supplementary artificial lighting in conjunction with side lighting.

THE ARTIFICIAL LIGHTING OF
FACTORY BUILDINGS

While it is generally difficult to consider the natural lighting of a factory in terms other than that of the building lighting, the artificial lighting, being more flexible in quantity and positioning, can be designed from the start independently as building lighting and as work lighting.

The building lighting should provide a general uniform level of illumination, and a distribution of brightness around the area, which will satisfy all requirements of amenity. In addition to this building lighting, local lighting on the work, either in the form of individual fittings or of additional bulk lighting, may also be provided where work of special visual difficulty, or work which demands careful visual attention, is to be undertaken in the building.

CHOICE OF LIGHT SOURCE

Very little can be done about the design of the artificial lighting until a decision has been made regarding the choice of the light source. There are various forms of light source available for use in factories, but of these the fluorescent tubular lamp in its various forms is the only light source worthy of serious consideration. Incandescent filament lamps are inefficient.

New forms of colour-corrected mercury discharge lamps, the mercury halide lamps, may soon become widely used. The colour rendering which they give is a great improvement on the discharge lamps which they are superseding. Their use can certainly be recommended in areas where their advantages of compactness and high luminous efficiency are of special value. Further developments in lighting fittings for use with these lamps will make them a potent rival to the fluorescent tube for general factory lighting, on the score of greater ease of maintenance because of their more compact size. Fittings for these lamps must however, be designed to limit glare discomfort, in order to meet the I.E.S. Glare Index limits for factory areas.

Tubular fluorescent lamps are available either in the form of hot cathode lamps (the common 'fluorescent tube') and also, much less frequently seen, the cold cathode tube.

Hot-cathode fluorescent lighting has been discussed in detail in Chapter 5. The cold-cathode tube has the advantage of high reliability and very long life, and although its initial cost is high, it is worth consideration where maintenance is unusually difficult or expensive.

CHOICE OF LIGHTING FITTINGS

The choice of lighting fittings to be used in a factory will depend more upon the Glare Index requirements and upon the ease of maintenance, than upon any other factors. While the lighting industry recognises the importance of glare elimination in factory lighting design, the design of factory lighting fittings is still not fully up to date with these ideas. As a result, there is insufficient choice in the type of lighting equipment for use in a factory, the choice lying between similar types made by different firms, rather than between radically different types of unit. It is therefore not easy to meet the recommendations of the Illuminating Engineering Society regarding Glare Index limits. On the other hand, the much less stringent requirements of the Factories Act can easily be met, but they represent a minimum standard of thirty years ago.

Choice lies between totally enclosed fittings and open fittings so designed that the heat of the lamp creates an upward draught through the fitting which carries dirt with it and so reduces the amount which settles on the reflecting and transmitting surfaces of the fitting. The reflecting surfaces of a fitting must not only have a high initial reflection factor, but must also be capable of being maintained in this condition. White vitreous enamel and anodised aluminium surfaces are preferable to synthetic enamels for this reason, even though the initial reflecting properties of synthetic materials may be higher.

Transparent and diffusing plastic materials have to a large extent replaced glass in totally enclosed fittings, but the use of such plastic materials must be related to the danger of fire in inflammable situations. Totally enclosed fittings, however, while apparently easier to clean, have their difficulties, not the least of which arises from the fact that the light output of the tubular fluorescent lamp is very dependent upon the ambient temperature. Enclosed fittings in a factory may therefore get too hot and the light output may fall. An open ventilating fitting is preferable, and experience shows that it is on the whole easier to maintain to a light output not far from the initial value.

ARRANGEMENT OF FITTINGS

The building lighting in a factory is almost always arranged either in the form of high bay lighting in which individual units are arranged symmetrically at maximum height from the working plane, or continuous run lighting, in which fittings are placed end to end along the length or across the width of a factory. The high bay lighting system, when well engineered, gives a uniform distribution of illumination. It is equally suited to units of fluorescent fittings, or of colour corrected mercury lighting.

The continuous run system has two advantages. First, it can be closely integrated with the daylighting, the continuous run of fluorescent lighting fittings being set along the same lines as the continuous run of glazing in the roof, either to supplant it after daylight fades, or to act as a permanent supplement during daylight hours. The second advantage of continuous run lighting is that trunking can easily be used both for mounting the fittings and for distributing the electrical supply.

The third form of building lighting which deserves mention is the use of a luminous ceiling. Many factories, particularly those which demand very close control over the interior climate, have been provided with a fully luminous ceiling, consisting of a continuous surface of transmitting diffusing plastic material, behind which fluorescent lamps are arranged at regular intervals to give a uniform distribution of brightness as seen through the plastic ceiling. The advantage of this system is chiefly one of cleanliness. In situations where the elimination of dirt has to be taken to extreme lengths, this system is probably the best which can be devised for the provision of a high and uniform level of building lighting. Recommendations have been made by the Illuminating Engineering Society that the maximum brightness (luminance) of any luminous ceiling used in work places should not exceed 150 foot lamberts. Provided this limit is scrupulously observed, a luminous ceiling is a solution to the problem of lighting super-clean work places. Experience has shown, however, that a luminous ceiling must be supplemented by local lighting on the work, because the bright ceiling is a centre of visual attraction which, in the absence of local emphasis lighting on the work, can cause tiredness and discomfort.

FLICKER AND STROBOSCOPIC EFFECTS

One of the main reasons for condemning uncorrected discharge lamp lighting is the fact that flicker and stroboscopic effects are marked, and can cause discomfort and even danger in the judgement of speed of rotating machinery. Modern fluorescent lighting gives less trouble on this score, but from the nature of the method of light production in the lamp, some residual flicker will always be present. This is true even in well engineered installations. In badly engineered installations lamp flicker may be acutely annoying. In addition, lamps which are nearing the end of their useful life tend to flicker, and it is not uncommon to go into a fluorescent lighting installation and be immediately aware of one or two lamps which clearly need replacing.

In some situations it may be necessary to take additional steps to eliminate flicker. One common procedure is to use twin-lamp lighting fittings, one lamp running on an inductive ballast, and the other on a capacitative ballast. This system causes the light emission of the two lamps to be out of phase and so for the peak light output in the one to fill the gap in the light output of the other. An alternative system is to wire the individuals in a triplet of lamps on the three phases of the supply. In some factories this system may also have advantages in the distribution arrangements.

DISTRIBUTION OF POWER FOR LIGHTING

The distribution of power for the building lighting is greatly assisted by the use of trunking. Trunking in the form of hollow aluminium or steel box structure can be joined and extended relatively cheaply and special devices are provided to enable the trunking to be supported from the roof structure at wide intervals. Lighting fittings can be attached at any necessary interval along the length and wiring up of fittings is simple. The trunking carries the wiring and the control gear for the lamps as well as for other services. It also has the advantage that the lighting installation can be supplemented at a later stage with little or no disturbance of the existing fittings and wiring.

Low-level trunking can also be used for the supplies for local lighting. This is particularly the case where the working areas are arranged in long assembly lines, as is very common in many factories. The electrical supplies for special directional lights can also be carried in low-level trunking although some additional method of fixing the local directional lighting will almost certainly be necessary.

INTEGRATED DAYLIGHT AND ARTIFICIAL LIGHT

It may often be advantageous to design a factory with mixed daylight and artificial lighting. In certain environments, particularly in cold exposed areas, the loss of internal heat through windows may balance the cost of artificial lighting, yet a totally windowless environment may be undesirable for psychological reasons.

An integrated lighting system in a factory can be considered in the form either of side windows supplemented by roof lighting from artificial sources only, or of side windows supplemented by partial roof daylighting and artificial lighting, or of a system where the top lighting is still dominantly from the windows, but which is provided with an artificial supplement for use on dull days.

The problem of supplementing, by artificial illumination, the lighting from a side window in a factory is precisely the same as in any other side-lit interior, and has been discussed in Chapter 6.

The provision of basically artificial roof lighting, or of supplementing daylight from the roof, raises somewhat different problems. The requirement is not so much for a balance of brightness, but for an installation which produces a uniform distribution of illumination on the working plane, and a satisfactory level of environmental brightness, in relation to the brightness produced by the side windows. In a large factory area, for example, where the floor area may be more than twenty times the square of the height of the room, the brightness distribution produced by the windows will have little effect in the centre of the room. If the sky is seen through the window, which will be the case in an unobstructed site, sky brightness will to some extent affect adaptation, and will certainly act as a distraction if the interior illumination is not of a high level. The aim should be to provide an illumination both on the working plane and on the surroundings, which, taken with the reflecting properties of the interior surfaces, will produce an environmental brightness not less than one-twentieth of that of the visible sky. In practice this means that general lighting, giving between 500 and 1000 lux (50 and 100 lm/ft²) is necessary.

In a single-storey factory, where some roof daylighting can be provided, the level of supplementary artificial illumination can be less. If it is to be used as a permanent supplement, it should be of such a level that, taken together with the available daylight from the roof windows, a level of the same order, i.e. 500 to 1000 lux (50 to 100 lm/ft²) will be available on average days.

The success of an integrated installation of this kind depends at least as much on the way in which it is designed with the daylight, as upon the actual level of illumination provided on the working plane. If the lighting is in the form of continuous runs, it can be linked to the design of the runs of glazing, for example by being placed beneath or to one side of the glazing or integrated to form part of the structure. There is wide scope for intelligent design by the architect and with skill it is possible to organise the lighting on the various floors of a multi-storey factory so that those who work on the lower floors feel no sense of deprivation of daylight relative to those who work on the top floor with direct access to the light of the sky through the roof.

Photo-electric control of the supplementary lighting, to

17

maintain a constant level of illumination in spite of changes in available sky light with time of day, cloudiness, etc. is a feasible proposition in a factory, but the control unit *must* operate a dimming system and not a switching system, and the rate of dimming *must* be related to the known characteristics of visual adaptation, otherwise the impression will continuously be given that the apparatus is out of order, bringing on the lights when they are not necessary and dimming them down prematurely. Such a controlled dimming system can be devised, but the cost of the installation must be balanced against the saving in electrical power costs during the total life of the installation.

DEPRECIATION AND MAINTENANCE

Industrial buildings are subject to serious depreciation in terms of lighting, whether of natural light through windows or of artificial light. Windows become dirty and lose their transmitting properties. Walls, floors, and ceilings become begrimed, and unless they are regularly cleaned and decorated lose their ability to provide a reflected component of light. Artificial lighting fittings become dirty and the light output of lamps falls.

It is possible by regular maintenance of windows and internal surfaces to ensure that the daylight level does not fall to less than 80% of its initial value provided the cleaning is sufficiently frequent for the degree of atmospheric pollution characteristic of the locality and of the process in the factory. It is advocated that provided such a cleaning schedule can be maintained, the calculation of the initial daylighting should allow for this depreciation. The average level of lighting can be taken to be 90% of the initial, if the final level before re-cleaning is 80%. The daylight calculation should therefore take into account this depreciation factor of 0·9, and windows should be made correspondingly larger, internal reflection factors correspondingly higher so that combined, the initial Daylight Factor will be 1·1 times the planned value.

It is not advocated that allowance be made for depreciation

greater than this level. If due to difficulties of access, roof lighting cannot be maintained clean, or washing and cleaning of internal surfaces is impossible and therefore it is known that the lighting level will fall below 80% of the initial, it is recommended that the deficit be made up by artificial lighting. Clearly, if a large depreciation factor is built into the daylight calculations, very large windows will result, with the consequent expensive loss of internal heat, a greater liability to sky glare through the greater area of window, and the probability of downdraughts in very cold weather. In many areas it will be found that the extra cost of the artificial lighting is no greater than the cost of the losses due to excessive glazing.

Regular cleaning and relamping of artificial lighting fittings is essential. By suitable choice of fittings, maintenance can be reduced to a minimum, and the open top 'self-cleaning' type of fitting is advocated. Such fittings can be cleaned *in situ*, but this will have to be done with the minimum interruption of the work. There is a great deal to be said for the maintenance organisation to hold a 'bank' of clean fittings, which are brought to the factory floor, the dirty fittings taken down and replaced by clean fittings, and taken away to the maintenance bay for thorough cleaning away from the production areas. In this way cleaning can be more thoroughly effected, and the interruption to production is kept to a minimum. Whereas solid reflector fittings need cleaning every two or three months in a normal factory, open top fittings probably require cleaning only once every twelve months. The maintenance schedule will, of course, depend upon the degree of atmospheric pollution. Reference should be made to the Illuminating Engineering Society's Technical Report No. 7 on the maintenance of lighting.

It is usually economical to relamp fittings by a system of group replacement. If lamps are accessible, individual lamp replacement may be a reasonable proposition, but with high bay lighting or with fluorescent lighting of any type, where the services of a skilled electrician are usually necessary for the lamp replacement, labour costs will be high and individual replacement far more expensive than group replacement. Group replacement can only apply to a group of lamps which are all switched on and off together, because otherwise the lamp life in the group will not be the same. The usual arrangement for a group replacement schedule is to plan to replace all lamps at the point at which, by the recognised 'survivor curve' for the particular type of lamp, 90% of the lamps would have been expected to fail. In practice the schedule will be linked to the type of lamp in use and the duration of shift working in the factory. For example, for single shift working, group replacement of fluorescent lamps might be linked to an 18-month schedule. For double shift working lamps might be replaced annually, and for all round the clock working group replacement would be necessary approximately every 6–8 months. The schedule cannot be worked out by simple arithmetic because of the complexity of the many factors involved. Detailed re-lamping schedules have been worked out for different situations, related to lamp costs and the cost of electrical power, duration of working and labour costs for lamp replacement, and these schedules should be consulted if group replacement is contemplated. The schedules given above, for 18-month, 12-month, and 6- to 8-month replacement periods for single, double, and three-shift working should be regarded as approximate only, serving to give only a general idea of the probable frequency of group replacement under average circumstances.

LIGHTING IN RELATION TO THERMAL COMFORT

It is not possible to design the lighting of factories without reference to thermal comfort conditions. The degree of glazing in a factory roof has a profound effect upon the heat insulation of the roof, and therefore in the amount of internal heat loss during winter, the discomfort from cold downdraughts, and in summer, thermal discomfort arising from unwanted solar heat penetration.

Both on the score of winter heat loss and summer heat gain, it has often been advocated that factory roof glazing should be avoided, the building lighting provided by artificial means, and side windows alone being installed for the purpose of giving a view outside rather than to contribute much to the internal illumination.

Decisions on these matters are difficult, because they depend on so many uncertain factors. Among these factors can be included the local requirements of the process itself, whether the building has to be controlled within close thermal and humidity limits, whether dust must be rigidly excluded, and other factors which indicate in favour of a fully insulated building; whether experience shows that the work people wish to have natural daylight, experience which might be obtained from labour turnover difficulties in neighbouring factories ill-provided with daylight; whether the local cost of electrical power is sufficiently low to make total artificial light less expensive than the design and maintenance costs of roof lighting, costs which include the cleaning of windows; and many other factors which vary from one locality to another, from one process to another, and indeed, which are changing steadily as time goes on, with improved luminous efficiency of lamps, with increasing labour costs for building maintenance and with demands by people for comfort conditions far higher than those ever contemplated in industrial environments in the past.

Summer discomfort due to unwanted solar heat penetration has two effects, one the rise of temperature inside a factory due to the total solar heat gain, and the other the local discomfort arising from uncontrolled sunshine falling upon people compelled to work in a fixed position. The second can be mitigated by the provision of roof glazing which diffuses the sunlight suitably. The first can be reduced in severity by the use of selective heat-absorbing glazing materials or by the use of suitable external screens, which control the admission of sunlight but which interfere as little as possible with the admission of diffuse sky light.

The use of diffusing roof glazing can have little effect upon the internal level of daylight from the sky, because such materials can be obtained with a high degree of diffusion and a relatively slight effect on the total transmission to diffuse light. The disadvantage of such materials is that, when illuminated by direct sunlight, their brightness is well beyond the glare comfort limits. Provided, however, these windows are out of the direct view of the worker, he may experience less annoyance and irritation with this glare discomfort than he would from the thermal discomfort of a direct beam of sunlight. The use of special glazing which incorporates both a heat absorbing layer and a diffusing layer has advantages which experience suggests are greater than would be predicted by a direct calculation based on theoretical considerations. On the other hand, such materials are very expensive and their installation in the roof glazing system of a factory might well prove, taken together with maintenance costs, demonstrably more costly than the insulation of the roof and the provision of a high standard of well designed artificial lighting.

The psychological factors associated with the desire for daylight have by no means been resolved. A designer will forego daylight at his peril, yet there is accumulating sufficient evidence to show that, if the factory as a whole is well designed, if care is taken with the use of colour and light reflecting surfaces in the interior to provide interest and sparkle, and if the artificial lighting is designed with skill and attention to detail in matters of glare discomfort and brightness distribution, work people will accept artificial lighting over long periods of the day, provided that sufficient window area is installed to give them an adequate visual access to the world outside.

11
Residential Buildings

Few of the recent advances in the practice of lighting have found their way into the buildings in which we spend the greater part of our lives—our homes. Lighting in the home is under the householder's control, and in the absence of education in lighting he continues to do what he and his parents have always done, or else he, or his wife, copies or adapts some lighting 'gimmick' which has appeared in a glossy magazine or in the house of a local Jones.

Responsibility for this state of things is evenly divided. Little or no education in lighting is given in school, or even in courses of further education in home economics. Even if it were, however, the application of the knowledge so gained would await the penetration of this knowledge to the 'designers' of the great majority of homes. While the private owner who can afford a good designer for his house may be much better served, the council tenant and the private mortgagee in their small dwellings too often find that their ingenuity is taxed to the limit to overcome the constraints which result from the awkward placing of windows without apparent regard to the only possible positions of essential furniture and likely activities in the room, or from the single central ceiling point and the single wall 'power socket' which has to serve the electric fire, the television, the iron, and the blanket as well as the lighting. It is not only in small houses that these limitations apply. Even in larger dwellings, including 'modern town apartments', to have more than three wall outlets is a luxury which is given star billing in an estate agent's blurb.

There is no rational explanation for this state of affairs, particularly in an age when care and money are rightly spent on other aspects of house design. Even small householders respond willingly to the enticements of central heating salesmen and lay out large sums of borrowed capital for the improvement of their home heating, and they spend equally large sums for sophisticated electrical equipment from television to deep freezes.

Improvement might well start with the architect who designs a house. There is a hope that if the windows can be planned rationally, and be seen so by the occupants, and if a certain limited freedom can also be given to the designer to build into the house some essential artificial lighting, such as that for the stairs and other circulation spaces, the owner may, with the help of some education in lighting, get the message and complete the job to his own taste and preference, but with the essential principles of good lighting satisfied.

THE PROBLEMS OF HOME LIGHTING DESIGN

It is certainly true that the designer of lighting in the home is faced with as complex a set of problems as he will meet in more expensive and heavily serviced types of buildings. The home contains a great variety of human activities which embrace a very wide range of visual tasks and concurrent emotional situations; in fact, the larger the home, the easier it will be to design the lighting, since there will be a better chance of confining different activities to different rooms, and so of providing for these different activities just the lighting which they should have. In the small house the same 'living room' has to provide for highly exacting visual tasks, such as darning, often done by elderly people whose eyesight is not what it was, and for other activities such as a teen-age party for which the absence of light is a prerequisite. Certainly in the same room one may be darning, one may be struggling with homework, and others may be watching television, all at the same time. Again, the emotional situations which the lighting will have to cater for will range from a young boy's rumbustious birthday party to his elder sister's first steps in courtship. All these activities happen in an area rarely more than 200 sq. ft.

The only solution lies in flexibility and adjustability, both of the lighting and of the activities. Difficult visual work can be taken to the window, or a table or standard lamp can be brought to the armchair or homework table, provided that the room can be planned and the power outlets supplied to allow this to happen easily and without dangerous trailing wires. The same problems and solutions apply to the other rooms of the house—kitchen and bedrooms, and study and playrooms if they exist —because all these have a wide range of visual activities also.

DAYLIGHT

Daylight, with all it implies, is essential to the home. No one has yet suggested a 'windowless' house, for everyone appreciates the stimulus and variety that the daily shift and seasonal change of the sun and sky will give both to the house and the scene outside. Everyone takes keen pleasure in basking in the rare February sun in a chair by a well-placed window, or in a carefully composed view of the garden through the window, or in watching the slow fading of light in a room on autumn afternoons. It is through windows that the vivid long-remembered childhood experiences come, of watching from the shelter of the home nature's more dramatic moments—the half enjoyed, half feared thunderstorms, the torrential downpours, trees bending in high winds, and the wished-for fall of snow which then magically suffuses the house with fresh white light.

The sensitive designer responds to the need for all these emotions but he recognises that they do not just happen unless he attends to the details in the design and placing of the windows. A glimpse of the sun may be welcome in one's bedroom in the morning, but the need for intelligent orientation will be appreciated by any parent who has tried to settle a child on a summer evening in an overheated room with a west-facing window. Again, the desire for a view out must be reconciled with the privacy that most people wish to have from others seeing in. As with other types of buildings, the easy contact with the outdoors which large windows can afford must be balanced against the disadvantages of both the heat which will escape through them in cold weather, and the risk of over-heating during at least a part of the summer. These difficulties may be met by the cheaper and more effective double glazing that is now available, and by the intelligent use of window blinds which can be adjusted to meet the vagaries of our summers. In the longer term, the planting of deciduous trees can help to shade a house during summer while still allowing a more open prospect during winter. Planting and screen walls or fences can also be used to give an enclosed, low brightness view through a window which will create more comfortable conditions for seeing with less glare than a direct view of bright sky.

Opening lights should be arranged so that transoms do not come at sitting or standing eye levels, and uninterrupted glass between sill and door head height is preferable, at least for the major part of the window.

Official regulations and codes of practice affecting daylighting were first drawn up in the nineteenth century as a defence against the speculative builders of back-to-back housing in rapidly expanding towns. As one would expect, the regulations were mainly concerned with ensuring certain minimum standards of light and ventilation. Thus the London Building Acts require that all habitable rooms should have a glazed area of not less than one-tenth of the floor area, half of which must open, and that 'the top of the opening of each window in the topmost storey is at least 6 ft. 6 in. above the level of the floor, and in every other storey is at least 7 feet above the level of the floor'. Similar by-laws operate in other parts of Britain, although some are not as stringent as those in London. The consequence in terms of daylight is that a room with decorations of average reflectance which complies with these by-laws will have a minimum daylight factor of about 1%. The current British Standard Code of Practice on Daylighting makes more detailed recommendations:

Kitchens. A minimum 2% D.F. over half of the floor area, with a minimum of 50 sq. ft., the intention being that sink, cooker, and preparation area should fall within this zone.

Living Rooms. A minimum 1% D.F. over at least 75 sq. ft., penetrating not less than three-quarters the depth of the room.

Bedrooms. A minimum 0·5% D.F. over at least 60 sq. ft., and penetrating not less than three-quarters the depth of the room.

In the case of bedrooms, the by-law requirement for a window at least 1/10th of the floor area will in most circumstances give a minimum daylight factor higher than that in the Code. In practice, in order to avoid an undue sense of enclosure and to give a good view out, most architects will design windows rather larger than those required as a minimum by either the by-laws or the Code of Practice.

DAYLIGHT AND GLARE

The standards of daylighting quoted above assume that the average reflection factors of the main surfaces of the room are likely to be not more than
  Walls: 40%   Floor: 15%   Ceiling: 70%
In domestic rooms of normal size this would give a minimum internal reflected component of about 0·2%. This means that in most parts of the room the proportion of direct light to indirect light will be high, especially as one is dealing with relatively small rooms where the working positions are likely to be close to the windows. There is, therefore, a risk of harsh contrasts in brightness and of sky glare, especially if the occupants choose dark curtains or carpets. Observations confirm that sky glare is in fact the commonest fault of daylighting in houses, especially when there are windows in one wall only.

In rooms that can only be lit from one wall, all the means of reducing discomfort glare that have been discussed in Chapters

3 and 4 should be considered. For a comfortable gradation of light from outside to inside, use should be made of tapered glazing bars, splayed jambs and deep sills. Wide transoms and mullions and 'natural' dark wood frames should be avoided. The window wall should be painted white or a light colour. In blocks of maisonettes a window extending the full width of the external wall is often necessary to ensure a sufficient penetration of daylight, especially when overshadowed by projecting balconies. While this will remove the problem of dark areas of the internal surface of the external wall being seen against the sky through the window, thought must be given to where curtains will hang when drawn back. The artificial lighting should be arranged so that it can be used conveniently to supplement the daylight on dull days, for example, with a fitting over the dining table, which is often placed at the back either of the living-room or the dining-kitchen.

The need to consider possible furniture layouts in relation to window positions has already been mentioned in connection with the usability of the rooms, but it can also have an effect upon visual comfort. For example, a common shortcoming in bedroom planning is when the only workable position for the bed is with its foot toward the window. Then anyone sitting up in bed and reading—whether ill, or just with the Sunday papers— will face glare from the window and will have difficulty in getting direct daylight on the page. A layout in which the bed can be placed parallel to the window will help to correct this, and will give the patient the choice of lying on one side with a view out, or on the other with a dimmer, more restful view to the back of the room. Another detail is that one should avoid placing a fixed mirror, whether in bathrooms or bedrooms, so that it faces the window, for it will not be possible to make out either the stubble on one's chin, or the contour of one's lipstick when silhouetted against the reflection of a bright window behind. The intelligent use of rooms and layout of furniture will of course depend upon the occupant of the house as well as the designer.

WINDOW DESIGN—OTHER CONSIDERATIONS

Besides direct lighting considerations, the design of windows will have to take into account other matters which may have a bearing on their lighting performance. They will be expected to provide effective natural ventilation in a variety of weather conditions. They should of course be wind- and water-tight when closed, and should not admit water readily when providing fractional ventilation. On tall blocks of flats wind pressures may be very high at times, and wind and rain may move *up* the face of the building. Traditional flashing and weathering details need re-examining in the light of this.

The window should be easy for a housewife to clean, and its method of opening should not present hazards either to the very young or the elderly. Many people will experience a distressing feeling of vertigo when looking down from windows in tall blocks, because of the sense of a sheer drop from the sill. In a number of recent schemes narrow balconies running along outside each window have contributed to a feeling of security, as well as providing a real safety 'net', and have at the same time given access for maintenance. The projecting balcony may allow the window below to be taken up close to the ceiling without undue sky glare.

ARTIFICIAL LIGHTING

The commonest shortcomings in artificial lighting that may be found in modern houses as well as older buildings are, first, under-provision and, second, unsuitable and badly placed light fittings. There are some signs of a recognition of the importance of good lighting, but it is still too common for the lighting 'installation' in most rooms to be a single ceiling rose, centrally placed, with a foot or two of flex dangling from it, and one or two socket outlets in the walls. Provision in privately-built estates is often no better—and sometimes worse—than in Local Authority housing.

The official Parker Morris report (1961), which very creditably recommended a substantially high standard of domestic heating, dealt rather cursorily with lighting, relegating to an Appendix an important reference to the necessity of good lighting for ensuring safety in the home. This point cannot be stressed too often: too many accidents in the home—where the majority of all accidents occur—such as falls and scalds, are likely to have poor lighting as a contributory factor. However, the Parker Morris report does assist lighting—although indirectly—by recommending that increased numbers of socket outlets should be installed, and many of these can if correctly positioned be used for light fittings as well as for other types of electrical appliance. Many can usefully be twin sockets; the additional cost is little more than the price of a two-way adaptor, and in use much safer and more convenient. For a 5-bedroom house the report recommends 20 sockets as a desirable provision, including 5 in the living room, with 15 as a minimum. With a reasonable number in each room, sockets should be placed on opposite walls, on each side of the door or fireplace and with careful consideration of where lighting fittings, as well as other electrical appliances such as record players and television sets, may be placed, so as to avoid long trailing flexes. The report states that they had found that a widespread practice seemed to be to provide only the minimum number of outlets—usually six—for which the Electricity Board would supply a cable free. Any comment on this state of affairs would have to be related to the large sale of two- and three-way adaptors in electrical shops and 'dime stores'.

In the absence of sociometric surveys, simple observation suggests that every room in a home should be provided with at least six outlets, and the living-room with at least ten, of which five or six will be in permanent use (two or three standard and table lamps, television, radio, and/or record player, heater or convector) and the others available for other equipment. With the increasing use of electric devices in the kitchen ten or more outlets are needed here also. Only with such liberal provision can it be ensured that points are vacant for any auxiliary lighting which may be needed to supplement the permanent installation.

Overall levels of illumination are of less significance in domestic accommodation than in other types of buildings. However, it is clearly important that there should be an adequate level of illumination in rooms where visually detailed work is done over an appreciable area such as the kitchen or a workshop, and the I.E.S. Code recommends 200 lux (20 lm/ft²), which corresponds with the level of daylighting recommended in the B.S. Code of Practice. In other rooms the higher levels of illumination recommended for close work such as sewing and darning (700 lux) and sustained reading (300 lux) will be gained by local lighting. Well-placed, flexible, and glare free fittings will allow people to get as close to their work as its visual difficulty requires and, when combined with an adequate level of background brightness through the room as a whole, should allow even the most difficult jobs—Granny's darning has already been mentioned—to be done efficiently and in comfort without throwing out of balance the lighting in the rest of the room. The use of a number of fittings in each room will always be an advantage: ensuring cross lighting and thus an absence of harsh shadowing, giving flexibility in switching to meet a variety of uses. When integrated with the design and furnishing of the room as a whole, this approach to domestic lighting can create a pleasing variety with a relaxed and intimate character. It may be helpful to consider how these general ideas may be applied to the artificial lighting requirements of five of the main rooms in the normal dwelling: living-room, dining-room, kitchen, bedroom, and bathroom.

LIVING-ROOM

The main requirement of the average family living-room is that of versatility. It will be used for a wide range of occasions and activities—subdued or gay, individual or communal. The visual tasks will range from the children's homework and close work such as darning, to the casual reading of the evening paper and watching television—although one hopes that there will be another room, perhaps a well-lit bedroom, where homework may be taken to escape the overbearing presence of television.

Viewing television is itself a visual task that needs consideration, although it no longer requires the subdued lighting that was once necessary when picture brightness and contrast was limited. Unfortunately these conditions led to the general notion that television ought to be viewed in darkness or near-darkness, and this has persisted to the present day. Television receivers are now capable of producing a picture bright enough to be seen easily in ordinary domestic lighting, although lights which cause annoying reflections in the screen should be switched off. There should also be no lights close to the screen to cause glare or distraction.

The screen should be placed to one side of the window, or, if there is more than one window, it may be essential to draw the curtains to screen off light which causes reflections in the screen or direct sky glare in the line of sight. The level of the screen should be determined to give the greatest comfort. For most people this is on, or slightly below, eye level, but some elderly people, bothered by bifocal spectacles, prefer the screen to be a little higher.

The amount of light in the room is also a matter of personal comfort. In theory the average brightness of the picture should be just slightly brighter than the background wall against which it is viewed. Most people are happy with the room lighting given by one standard or table lamp (an average room luminance of about 10 asb, or 1 ft-L), lighting which also permits a non-viewer to read or sew by the lamp without interfering with the visual comfort of the rest. However, the table lamp should not be placed on top of the set. This is a current fashion, but it causes visual distraction whether recognised or not, and the background light which it provides is neither necessary or desirable.

The social occasion may rarely be thought of as a visual task but probably the most subtle domestic lighting problem is the conversation piece—simply talking with other members of the family or one's friends. The lighting should allow people to see each other easily, and one should not be forced to squint at faces with bright fittings just behind or in front of them. The lighting should be flattering, but not too revealing—this should be reserved for the moment of truth in the bathroom mirror. Our faces seem to look their best in lighting which in the main comes from one direction, but with some general diffused illumination to lighten shadows. A good position for the main light sources—which should not be too bright—is either level with or a little above sitting eye level, provided they are of completely opaque material at the sides. Strongly directional lighting from well above the face is not flattering. Recessed 'black hole' fittings (for which there has been a passing fashion can emphasise the bone structure of the face in a most ghoulish fashion, and should be used with discretion.

The most useful general purpose fitting for the living room is one with a large cylindrical shade and open top and bottom. This will give an 'up-and-down' distribution which will give both direct light on one's book or sewing and diffused indirect light reflected off the ceiling and upper walls. When mounted at or a little above seated eye level—that is, with the lamp at about 4 ft. above the floor—a shade with the right proportions will give a good spread of light sideways beneath the fitting and at the same time will give a cut-off to a direct view of the lamp when looked at by someone walking past. A common fault with such shades is that the supporting ring is placed too high, thus allowing the lamp to be seen too easily from above. A fault with drum shades made from an opaque material is that they are often lined with white and as a result are liable to give excessive brightness contrasts when the inside is seen against the outside. Translucent shades are appropriate for these fittings, provided they are not too bright—a luminance of between 10 and 100 ft-L will be found to be comfortable in most domestic interiors but a fitting of the higher luminance in this range should not be too much in direct view, for example, when watching television. Manufacturers should state the maximum wattage of lamp for each shade. Choice of colour will depend on the colour scheme in

18

the rest of the room. Shades in light, neutral colours will be found to be the most versatile, but if a definite hue is required, pinks, reds, and soft yellows will be kinder to complexions than blues, greens, and purples.

The type of fitting described above is close in many ways to the traditional standard lamp which has after all evolved by a process of trial and error over the fifty or so years of domestic electric lighting. With the lighter, more flexible furniture of to-day, which people can move around freely, the comparative instability of the standard lamp may be inconvenient, and the same type of shade mounted on a wide glass or pottery base placed on a low table may be preferable. A number of these 'up-and-down' fittings will usually be enough to light the activities in a normal-sized living-room, but some additional local or display lighting may be needed for special purposes. A writing bureau or a record player may conveniently have a small fitting incorporated as part of the original design. A lamp fitted beneath a trans-lucent glass shelf can set off a well-grown house plant—though it should not cause too much heat. A collection of glass or a special picture may require, on a small scale, individual display lighting. A picture can be picked out with a small floodlight but the source should be concealed or well-louvered, and placed so that there are not irritating reflections if the picture is covered with glass.

At times preferential lighting may be required over a larger area—a curtain drawn from wall to wall, a special wall paper, a group of pictures filling a wall. This can be done with 'pelmet' lighting, with linear tungsten or fluorescent lamps concealed behind a pelmet at ceiling level. A common fault is to place the lamps too close to the surface being illuminated so that there is a very bright patch at the top, often making a glaring contrast with the pelmet, and with the brightness falling away sharply lower down. To prevent the illumination on the wall being too un-balanced, the lamps should be at least 12 in. away from the wall, and the upper part of the wall shielded from direct light. An alternative way of bathing a vertical surface with light is to mount a row of floodlamps on—or recessed into—the ceiling at a distance of 2 or 3 ft. from the wall, pointing vertically at the floor. The manufacturers of the fittings chosen should be able to advise on the right spacing and distance from the wall to give an acceptably even illumination and to avoid excessive scalloping. It should be noted that the light will graze the wall surface, thus emphasising the texture of fair-faced brickwork (which may be what is wanted) or causing picture frames to cast heavy shadows (which may *not* be wanted).

The most satisfying form of lighting in a living-room will be one that is flexible, giving a good level of illumination without glare at the main sitting positions, with an interesting and varied distribution of brightness through the main surfaces in the room, but avoiding exaggerated contrasts or contrived effects which can soon become tiresome. Most of the lighting fittings will be chosen by the householder, so that the role of the architect is chiefly to provide the opportunities for the intelligent lighting of the room, with well-placed sockets and perhaps a certain amount of built-in background lighting which should be switched from conveniently close to the door.

DINING

Lighting the area where members of a family take their meals presents problems very different from the living-room. In place of a room that will be used in many different ways, the dining-room—or the dining area in a kitchen or living-room—has a single main function, although the settings for a weekday breakfast, the children's tea, and a special dinner party are not likely to be the same.

However, the principal visual tasks will be the same in each case. They are:

first, to permit everyone to see well, at least enough to be able to fillet the proverbial kipper easily, and to enable the middle-aged, who may not wish to put on their reading glasses for a social occasion, to be able to eat without apprehension, *second*, to make the food itself look attractive,

third, to put the setting of the meal—table, linen, cutlery, glass—in a good light, and

fourth, to be able to see the company around the table—family and friends—easily and in a flattering light.

The first three call for direct light on the table, preferably from small point sources which will bring out the colour and texture of the food, and make the glass, wine, and silver sparkle and glow. A good solution is a fitting (or fittings, if the table is a long one) placed so that the spread of direct light covers the table top but does not shine into the eyes of the diners (Plate 64). If it is hung high enough not to obscure the view of someone sitting opposite, there is the risk that the bright inner surface of the fitting will cause distracting glare, and it is better if the inner rim is of a dark colour. Translucent shades in coloured materials should be used with caution, as they can do unpleasant things to food—and one's guests' faces. Lighting faces is better done with light reflected from the table, and with the general background lighting which will also light the surroundings to a level against which people may be seen comfortably.

It may be the custom in the family for the Sunday roast to be carved at a side table, and some local lighting will make the task easier. It is probably better for this to be done with a movable table lamp from a socket outlet, rather than tie down the furnishing of the room with a wall-mounted fitting.

The traditional candle-lit table meets most requirements, but the modern fitting with imitation candle lamps does not, because of their very much higher luminance which causes strong glare. A hostess who delights in candle light can use the real thing in the secure knowledge that she is offending no principle of good lighting.

KITCHEN

The kitchen and its ancillaries such as a utility room are the places where the most exacting and sustained visual work in the home is carried out—cooking, washing-up, laundry, ironing. The well-being of the family, and certainly of the housewife, depend to no little extent on having good lighting in the kitchen. The cook has to be able to judge subtle differences in colour and texture of the food as it is cooking, often through the additional hazard of steam or at the bottom of deep pans.

Unlike most other rooms in the house the workplaces in the kitchen are usually fixed and are often, because of piped services, against a wall so that the cook will be working in her own light if she has to depend on a single window or a central light fitting. In fact, because the workplaces are fixed it is usually possible to plan them in the best relationship to the windows, and then to place light fittings that will throw light exactly where it is needed.

The window(s) in a kitchen will normally be placed to give the maximum light in the main working areas—sink, cooker, and preparation table or worktop. At the same time it should be remembered that the housewife spends a substantial portion of her working day in the kitchen and the need for a pleasant view out should be considered, either over a quiet garden or a busy street, or perhaps to keep an eye on young children at play. Other considerations in window design will be the need for adequate ventilation, and the placing of the window in relation to wall storage, which will be at a premium. The provision of some light reflective surfaces in the room, particularly the floor, will help to lighten shadows under work surfaces and in cupboards, and will make cleaning easier.

More than one artificial light fitting will be required. The correct method of planning the main working area of the kitchen is in the sequence: worktop—cooker—worktop—sink—worktop or drainer. This should be in a continuous run at between 2 ft. 9 in. to 3 ft. 0 in. high, preferably with an additional lower working surface, perhaps a loose table. This sequence may be arranged either in a straight line (for a small kitchen), or in a 'L' or 'U' layout. The whole of this worktop should be well lit, and a number of fittings (as many as three or four 100 watt fittings may be required) should be arranged along the run so that the housewife does not have to work in her own light at any point. The fittings may be recessed in to the ceiling, mounted behind pelmets or cooker hoods, fixed above or below wall cupboards. They should be designed for easy cleaning and maintenance. Tungsten filament lamps are still the commonest light source in kitchens. If fluorescent fittings are used, it should be remembered that many of the most efficient tubes do not render the colours of food accurately, particularly in turning the reds of meat brownish. Tubes which render reds accurately, such as the colour-matching types, should be chosen, preferably mixed with tungsten lamps to give sparkle. If mixed installations are used, it is better if the contrasting appearance of the lamps themselves are concealed as far as possible from the direct view, by louvres or behind pelmets. The need to be able to see easily into cupboards should be remembered when positioning fittings; some large cupboards may usefully have their own internal lighting.

BEDROOMS

Nowadays few bedrooms are used solely for sleeping, and bedroom lighting will normally have to take into account their use for relaxation, hobbies, and study.

The importance of the orientation of the room in relation to the sun, and of the placing of the window in relation to the position of the bed, has already been discussed. The most critical visual task is probably seeing oneself in a mirror, whether at a dressing table, or full-length view in a wall mirror. In either case the mirror should be placed adjacent to, or with its back to, a window. The former position will be better to prevent sky glare.

After dark the position in front of the mirror will also need to be well lit, and a light source on either side will give better viewing conditions than a single one overhead. The other local lighting needed will include a bedside light or lights, suitable screened or adjustable if it is a double room. Wall socket outlets in the right place will give greater flexibility in the choice and final position of fittings. The need to be able to see into cupboards should also be borne in mind when placing the main fittings. A desk light may be wanted if the room is to be used for study or hobbies. This range of local lighting should provide sufficient general lighting in most bedrooms, provided that some fittings are switched from by the door.

BATHROOMS

Lighting in the bathroom and the toilet has to be governed by electrical safety. There should be no lamp outlets at such a height or in such a position that an inexperienced person could touch both a live point and any earthed metallic fitment. Older bathrooms are unfortunately wired with outlets which do not meet these requirements, and it is probably only because electrical accidents are not more frequent than they are that some form of compulsory inspection is not introduced. Apart from the danger of exposed wires and points, which are dangerous everywhere but potentially lethal in the bathroom, the main danger is of lamp sockets at low level such that an unwary householder could change a faulty lamp while the point is still live, and in the process hold on to a tap or other earthed object.

For this reason, however desirable it may be on lighting grounds to have low-level lighting, for example by the mirror, for reasons of electrical safety the lighting should be mounted in the ceiling in special shielded sockets which make it difficult to insert the fingers when changing a lamp. Most bathrooms have white or light-coloured surfaces, and this enhances the internally reflected lighting, so that one lamp outlet, or two in a large room, can provide adequate illumination for the few visual tasks, like shaving, which require help from the lighting. Switching should be by ceiling-mounted, cord-operated switches, or similar safety devices.

The bathroom should never be equipped so that the mirror has to face the window, because the person using the mirror needs light on his face and not behind him. Otherwise he sees himself silhouetted darkly against the image of the bright window in the mirror.

There is no objection on lighting grounds to the use of fluorescent lighting in the bathroom, but some, in fact most, arrangements of fluorescent lighting call for stable earthing devices to ensure ready and reliable starting and running. In a steamy bathroom difficulty may arise, and although it is not insurmountable, it is best to be sure of the competence and skill of the electrical installer. No such problems arise with filament lighting.

OTHER AREAS

The lighting of other 'specialist' rooms or activities in the home can be derived from an application of the methods described in other chapters, and need not be discussed in detail here. However, one matter must be mentioned: the need for the careful lighting of stairs and landings so that all possible causes of accidents can be removed. Light fittings should be positioned so that the treads are illuminated clearly, and so that one does not have to look down into a glaring fitting when coming down stairs.

It is appreciated that many of the points discussed in this chapter are mainly under the control of the householder. Nevertheless the architect and lighting designer can do much to encourage good lighting in the home by planning carefully the positions of lighting points and socket outlets in relation to the likely furnishing and use of the rooms, and by considering the design of the fittings themselves in relation to the distribution of light and the cut-off to the bare lamp that is required. Beyond this, a straightforward explanation of these considerations to the incoming householder would help to ensure that the best use is made of the lighting facilities that have been provided.

DWELLINGS FOR OLD PEOPLE

The problems of lighting dwellings for old people are rather different from those for ordinary family houses and flats. It is a characteristic of old age that eyesight fails in a number of different ways. Old people do not see as well as young people even in good light, and in addition to suffering from some cloudiness or blurring of vision, they have difficulty both in adjusting their focus to objects of different distances and in adapting quickly to different brightness levels when passing from a light area to a dark area. Old people also have difficulties in interpreting visual information and in making decisions on the basis of what they see. It has often been reported, for example, that after an accident involving an elderly pedestrian, he may report that he saw the vehicle coming but was unable to change his intention of crossing the road in time to take avoiding action. In the home elderly people often make apparently stupid mistakes which could be explained if they had not seen a hazard, but it often proves that they did see it but were unable to interpret the visual information sufficiently quickly to take the appropriate action. There is good evidence that their processes of seeing and subsequent decision-making can be speeded up usefully if they have good lighting in their living quarters. The provisions which are made to assist the vision of normal sighted people are particularly important to the elderly. It is desirable to improve the contrasts in the visual task or between the task and the background where this is possible, for example, by suitable decoration in the home. It is desirable to have a strong light on the work but there must be good general lighting as well otherwise when the eyes look up from the work, there may be a temporary blackout due to the slowness of adaptation if the lighting in the rest of the room is not of adequate level.

Glare must be avoided because there is no doubt that most elderly people are more troubled by disabling glare (dazzle) than are younger people, while those who are suffering from the early stages of, for example, cataract, will be considerably handicapped by veiling brightness possibly before their condition has been diagnosed. Windows should therefore be provided with blinds or curtains which can be adjusted to cut out sky glare while still permitting good daylight to enter the room. At night lighting fittings should be so designed that there is no direct view of the lamp and even the bright parts of the fittings should be severely limited in brightness and area. Large bright diffusing globes and cylinders are therefore to be avoided.

It is possible to give considerable visual aid by the intelligent use of paint and colour for picking out objects which need to be seen easily and without effort, such as the position of light switches and power outlets, kitchen equipment, and other objects which must be seen and identified with certainty. The greatest care must be taken in the painting of steps and any other changes in floor level. Clearly such changes in floor level should be avoided wherever possible. The colours to be used on steps and risers will depend upon the direction of the dominant light. In general, however, it is better to paint the risers and the treads in contrasting reflectances so that the boundary can be clearly seen both when going up and when going down stairs. Doors and their surrounds should be painted in such a way that it is always possible to see clearly whether the door is open or closed, and door handles should be made visible by a contrasting brightness or colour.

The Joint Committee on Lighting and Vision of the Medical Research Council and the Building Research Board has made certain recommendations for the quantity of light to be provided in dwellings for old people. These recommendations in general demand a minimum daylight factor of not less than 1 % in all rooms except kitchens where it should be 2 %. Attention is drawn to the fact that a view window is desirable in all rooms but that such windows must be opened easily and it should not be necessary to lean over furniture, the kitchen sink or worse still the cooker, in order to get to the window to open it. In the living room provision should be made for the work to be taken to the window for better light and a bay window or corner window is recommended because old people often like to sit by windows, to see out. People who are confined to bed will also be grateful for a window with a view.

It is recommended that artificial lighting should be by filament lamps because of their easier maintenance, and the maintained level of lighting should be not less than 100 lux (10 lm/ft²). Local light for work is recommended, but there should be no trailing flex, and above all flexible leads on the floor must be avoided because they can cause stumbling. Power outlets should be not less than 3 ft. above floor level and an adequate number should be provided. Lighting should be provided with two-way switches at entrance doors, and in the bedroom from the bed position and the door.

Television should not be viewed in a dark room but in the general lighting of the room because screens are amply bright when the set is well adjusted. Unwanted reflections in the screens must, of course, be avoided. By day the television set is preferably to be placed on a wall adjacent to the window wall but turned away from the window to avoid reflections and lowered contrast.

Elderly people tend to feel guilty about the use of a great deal of artificial lighting. They grew up when artificial lighting was very expensive and they were trained to switch it off whenever it was not being used. This is, of course, a dangerous practice and there are strong recommendations that in local authority homes for old people, the cost of the lighting should be handled on the same basis as the central heating and other services. If the elderly person clearly understands that any savings which he may make in the lighting will be of no financial benefit to him, it is possible that he may take a more sensible attitude to the use of artificial lighting. Even so, many elderly people have strong consciences in matters of this kind and a great deal of education is necessary to make sure that they understand their own best interests.

Lighting in old people's homes should be related to the various visual aids which are now available to assist failing vision. Among these aids are special reading desks which incorporate a reading lamp and a magnifier. When well designed, such a device is of great help, but the lamp must be completely screened from the reader's eyes, and also from the magnifying lens, because otherwise the veiling and reflections will be so troublesome as to make the device useless. Designers of these visual aids sometimes appear to forget that one of the main causes of loss of vision in the elderly is this veiling effect due to scattered light in the eye.

For the same reason, special reading material is often of

greater value if printed in white letters on a dark ground. Many elderly people with senile cataract can read print of normal size when printed in this way, and find this much more convenient than the very heavy large-print books that the well-meaning provide for them. White print on a black ground can easily be prepared from normal print by a simple photographic process. Cataract patients can often read such print in average lighting, but may be greatly assisted by additional light well screened from their eyes.

12

Buildings in the Tropics

The best buildings in tropical countries make use of the indigenous architectural tradition, extending it and developing it in the light of modern technology where required. The worst buildings import the traditions of colonising countries with little or no intelligence applied even to modifying this alien tradition to the totally different climatic situation. The growth of building construction in tropical countries has made it necessary to develop a new technology for the lighting of such buildings both by natural means and with artificial lighting, to take into account the special climate and the local traditions while still maintaining the fundamental principles of good lighting which apply throughout the world.

It has been suggested that there are two basic climates which are characteristic of tropical regions, both of which are significantly different from the overcast, cloudy climate of maritime north-west Europe, upon which most current lighting technology is based. These two tropical climates are the hot, dry, sunny climate, dominated by the unobscured sun and the cloudless, clear blue sky, and on the other hand the hot, humid climate with a high variability of cloudiness together with frequent seasonal sunshine.

The hot, dry climate is a relatively stable condition which lends itself to the development of a lighting technology. The sky under these conditions is often of very low luminance, insufficiently bright to act as the principal source of interior illumination. On the other hand, the uninterrupted sunlight lights up all surfaces upon which it falls, the ground, building façades, etc.

and these surfaces may well have a brightness (luminance) well in excess of that of the sky, and so will act as the main sources of natural lighting for the interiors of buildings. Lighting technology in the hot, dry climates makes use of this reflected sunlight as the major source of working illumination. The problem then becomes one of designing a window which will admit such reflected sunlight while at the same time excluding unwanted solar heat radiation and excessive glare. The architectural tradition in these areas is designed, consciously or unconsciously, around such a concept. It makes use of high windows, sunbreaks and other forms of screen, and above all arranges buildings around shaded courtyards with the use of wherever possible trees and vegetation to act as shading.

The hot, humid climate, on the other hand, is characterised by skies of very high brightness. Consequently the view of the sky needs to be screened from all internal viewing angles, and in addition windows must be placed and designed to permit the maximum air movement indoors, taking such advantage as is possible of any prevailing breeze by orientating windows suitably. The vernacular architecture in these areas follows this pattern, with overhanging eaves and deep verandas.

It is impossible to consider the lighting in buildings in the tropics without considering the heating problem at the same time. Windows often have to be designed not so much around the admission of daylight as around the elimination of unwanted heat. Yet at the same time if windows, designed for the exclusion of solar heat, admit as a result insufficient working light into the room, recourse cannot be made to artificial lighting because this in itself will introduce additional heat and so no advantage will be gained. It must also be borne in mind that only in a few areas in the tropics are excessively uncomfortable thermal conditions experienced throughout the year. Some areas, for example, the highlands of central-southern Africa (Rhodesia and Zambia), although lying within the tropic of Capricorn, nevertheless experience 'winter' conditions which can be uncomfortably cool and which may occasionally call for artificial heating. The warming effects of the sun under such conditions are welcomed. Consequently window screening devices which completely exclude solar radiation may not be desirable, and it may be preferable to introduce some form of control which permits the necessary adjustment. In other tropical areas, for example Singapore and Indonesia generally, the climate is characterised by a fairly continuous uniform temperature of the order of 25°C (78°F) which fluctuates very little throughout the year but which is accompanied by high humidity causing discomfort which can only be alleviated by constant air movement. It is therefore evident that to attempt to deal briefly with the whole problem of lighting throughout the tropical regions can do no more than to indicate that there are special problems quite different from those characteristic of north-west Europe, but which can nevertheless be handled by an informed application of basic principles.

Although no systematic study has been published of necessary levels of illumination for the interiors of buildings in the tropics, it appears to be agreed that an interior illumination of the order of 500 lux (50 lm/ft²) rising to 1000 lux (100 lm/ft²) is satisfactory as a general working light. This is somewhat higher than that at present characteristic of conditions in Great Britain but can be justified if necessary on the grounds that since it is much brighter outside, interiors should also be brighter to counteract unfavourable adaptation effects as one moves in and out of buildings. It is possible to obtain interior illumination levels of this order by judicious design making use of reflected sunlight in a hot, dry climate or of skylight in a hot, humid climate.

DESIGN FOR A HOT, DRY CLIMATE

The design of windows in tropical buildings must be related not only to the somewhat different demands for lighting levels, however, but also to the different amounts of light available outdoors. Available daylight is related to the latitude directly, and so it might be expected that the same size of window which provides adequate interior daylight in temperate latitudes will provide the desired higher levels of interior light from the brighter skies of the tropics. This is in general true, but the situation is complicated by the need for very different design of windows for tropical regions, particularly the introduction of permanent sunbreaks and louvres which greatly modify the light-penetrating properties of a window.

In a sunny climate, the light which penetrates to the interior of a building is composed of three main components, the direct skylight as in a temperate area, and then in addition the light reflected from the ground and the light reflected from the vertical façades of other buildings and other vertical surfaces. These latter two components, which under an overcast sky are almost negligible, can constitute the major sources of light in a sunny, dry climate. It has been demonstrated experimentally that these three components change relatively during the day in such a way that the sum of their contributions to the internal daylight in a room remains more or less constant throughout the period during which the room is used for work.

The calculations for the design of windows can best be handled on this assumption, and it will then be found that the interior illumination depends upon only three chief variables, the glazed area of the window expressed as a fraction of the floor area, the reflectances of the internal room surfaces, and the distance of the reference point from the window. Nomograms have been prepared by Plant at University College, London, which relate these three factors and which permit the computation of the internal illumination for windows of different sizes with and without sunbreak systems placed over the window. The nomograms equally permit the size and position of windows to be predicted in order to obtain a given level of interior light (Fig. 22).

The reflectances of the interior room surfaces are important just as in temperate climates. In temperate climates, however, the main light flux comes from the sky and is first reflected from the floor near the window. In the sunny climate the light flux

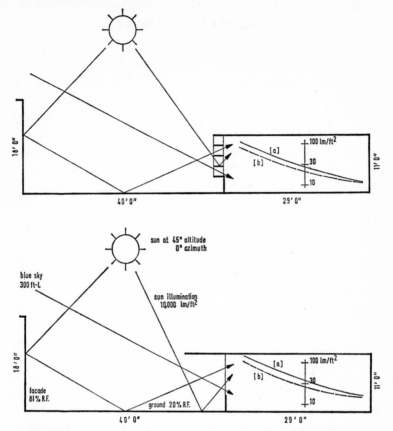

Fig. 22. Daylighting in hot dry climate by reflected sunlight (after C. G. Plant). In hot countries where continuous sun and blue skies are the rule, daylighting techniques based on the overcast European sky cannot be applied. Windows in traditional buildings in these countries are often placed to face away from the sun and rooms rely for their illumination on light reflected off the ground or buildings opposite. New buildings can be designed on the same principles. The diagrams show the illumination levels that will be found in typical rooms in which the windows are oriented and screened so that no direct sun may enter the room. The room in (a) has a deep canopy. In (b) the screening is by an egg-crate louver with a 45° cut-off. Of the two illumination curves shown: (a) is that with a window of 20% of the floor area and walls of 48% reflectance, and (b) is that with a window of 10% of the floor area and walls of 81% reflectance. In practice the size of the windows and the method of shading them will be related also to the problems of ventilation, and of the heat gain from reflected sunlight. Interior reflectances will be chosen in relation to those outside in order to reduce glare.

comes from all directions, and probably least from the sky, and so there is no need to give preferential consideration to the floor reflectance. Care has to be taken, however, to ensure that reflectances are not so high that under some conditions there will be discomfort from glare. Full white should be avoided on all those surfaces which receive the most light, and reflectances should never exceed 70% (Munsell Value 9). The level of illumination in the building interior is highly dependent upon the interior reflectances because direct light, whether from the sky or from external reflecting surfaces, is usually restricted by the existence of sunbreaks or louvres over the window. Reflectances of walls and floor should be in the region of 50–70% (Munsell Value 7·5–9).

Within these reflectance limits a wide range of colour treatment is possible, all colours within the Munsell Value range 7·5–9 being available. It is also permissible to use small areas of lower reflectance and therefore of strong colour since the inter-reflecting properties of a room are dependent upon the average reflectance of all the surfaces in the room. A relatively small area in terms of the fraction which it occupies of the total surface area may nevertheless have a dominant effect on the visual environment because of its position in the field of view. It is desirable to point this out because it has sometimes been suggested that the above recommendations for interior reflectances in tropical buildings seriously limit the colour treatment which can be employed. This is clearly not the case.

A building design which relies upon reflected sunlight as the main source of illumination depends less upon orientation for the quantity of interior illumination than upon the area of fenestration, the reflectances of the internal surfaces in the room, and the reflectances of the sunbreaks.

THE HOT, HUMID CLIMATE

Tropical areas where the climate is governed by the nearness of the sea are characterised by skies of variable cloudiness but without the frequency of fully overcast conditions characteristic of north-west maritime Europe. It is consequently not feasible to base the design of the interior natural lighting upon the international standard overcast sky (Chapter 4) as in Great Britain. Precise calculation of the daylighting is, in fact, quite impossible because of the extreme variability of the exterior illumination. The convention has therefore been proposed to base such design upon a sky of uniform luminance distribution, but to use very simple methods of calculation since precision methods would be misplaced.

Interior daylighting in the hot, humid climate is best estimated by a method devised in the United States and adopted by the American Illuminating Engineering Society (Recommended Practice of Daylighting, American I.E.S., 1962). The daylight factor is given by the formula:

$$D = FA_gK_uK_m$$

where F is a window factor giving the ratio of the vertical illumination on the window from the sky and from the ground to the total outdoor illumination on a horizontal plane, A_g is the total area of actual glazing, K_m a maintenance coefficient and K_u a coefficient of utilisation.

The American I.E.S. publishes tables of coefficient of utilisation for positions at different points in the room. Other tables take account of the effect of using horizontal sunbreaks tilted at 45° and for diffusing blinds or curtains.

Window design in the hot, humid climate is not, however, similar to that for the overcast sky in temperate regions. The sky will usually be too bright for visual comfort, and some form of permanent screening device is necessary. This permanent screening may be in the form of louvres over the upper parts of the window, which may be fixed or adjustable. The brightness of the ground and opposing façades may also be beyond the comfort limits, particularly when they receive direct sunlight or high illumination from a very bright sky. Practice in the United States, for example, makes use of glass of low transmission for the *lower* parts of the window. Low transmission glass can take the form of neutral glass, i.e. glass which has no selective colour transmission. Special glass with anti-thermal properties is sometimes advocated because this glass is selectively transmitting to visible radiation. Such anti-thermal glasses which operate by absorbing heat radiation are not entirely satisfactory, however, because this absorbed radiation will raise the temperature of the glass and the window will consequently act as a heat radiator. This is particularly troublesome if people are not able to move away from the window. Selectively absorbent glasses of this type also tend to be coloured, and if such a tinted glass is not used to glaze the whole of all the windows, unfavourable colour comparisons are made between the treated and untreated areas of a window. If all the windows are glazed with anti-thermal glass, the eye tends to adapt to the slightly greenish colour and after a few moments no colour distortion is noticed until the window is opened or people go out into the open again and they are then bothered for a few moments by the apparent pinkishness of the world outside.

New types of selectively reflecting glass are being developed which, though at the moment expensive, may in the event prove to be the best solution to the solar radiation problem. Since these glasses are designed to reflect the unwanted heat radiation while transmitting the wanted light radiation, the thermal build up in the glass itself is minimal. They do have an interesting side effect, however. Their higher reflectance results in reflected images of the interior when the artificial lighting is on and the daylight outside is on the low side. People see themselves reflected in the window against the visible background of the view outside, and the incongruous scale effects—walking about among the skyscrapers—is distressing and has been the cause of comment and complaint. It may be, however, that this is an effect to which people learn to adapt.

The design of windows for the hot, humid climate is, however, as much a matter for the provision of ventilation and air movement as for lighting, unless full air-conditioning is provided. Arrangements are necessary to open windows completely to allow air to circulate while at the same time providing means for exercising protection from driving rain, or, at some periods of the year, dust storms. The cost of special heat absorbing or reflecting glasses in such openable windows should be weighed up with some care as a consequence. It may well be argued that such glasses can be afforded only where there is also complete air-conditioning.

THE CONTROL OF SUNLIGHT

One of the main problems in tropical building is the exclusion of sunlight, and this has often to take priority over the lighting. Solar exclusion devices must be placed outside the window if they are to have the maximum effect in excluding unwanted heat radiation, although they are equally effective inside so far as lighting is concerned.

The design of sunbreaks demands skill in solar geometry, but there are now many aids available which simplify the design considerably. The Phillips Shadow-angle Protractor is one of the most useful of these devices. There are also sunlight-indicator diagrams published by the French Centre Scientifique et Technique du Bâtiment which cover all latitudes at 2° intervals, which can be used directly with architects' perspective projections (the Phillips protractor requires the preparation of a stereographic projection).

Solar exclusion devices take many forms, including the use of fixed vertical and horizontal louvers, or of honeycomb louvers. Horizontal louvers can be designed to exclude the sun completely, but to some people the severe restriction on the view which results negatives the value of the window. Vertical louvers permit a view outside, and many people consider that the interference with the view which they cause is less irritating than the effect of horizontal louvers. The vertical louvers in the windows of the American Academy of Sciences building in Washington D.C. are motor operated to move round with the sun, so that the view changes also, but it is unusual to provide such adjustment, the louvers instead being set to the angle which gives the best protection for the particular orientation of the window. This angle can be determined with the aid of the shadow protractor. (See *Daylight:* Hopkinson, Petherbridge and Longmore, Heinemann 1966; and *Sunlight in Buildings:* ed. Hopkinson, Bouwcentrum, Rotterdam, 1967.)

ARTIFICIAL LIGHTING IN TROPICAL BUILDINGS

The development of the fluorescent lamp is of considerable advantage in planning the artificial lighting in tropical buildings. The total amount of heat developed (by radiation and convection) in an installation from high efficiency fluorescent lighting is of the order of only one-quarter of that developed from a comparable installation of filament lighting. More important, however, is the fact that it is possible to integrate the fluorescent lighting with the building structure and to arrange that a current of air is drawn out from the room over the fluorescent fittings and this heated air directed out of the room. The only heating effect which remains is then the radiant effect which, though not negligible, is unlikely to cause great concern. During the cool season, use can be made of the convected heat if required and it can supply the whole of the heating necessary for the room.

The colour of fluorescent lighting to be chosen for tropical buildings appears to be related in some way to the basic culture of the inhabitants. It has been stated, on the authority of the sales organisations of lamp manufacturers, that in India, for example, the Indians themselves prefer a different colour of fluorescent lamp in their offices to that chosen by the domiciled Europeans. The reasons for this can only be guessed, although it has been stated that the Europeans like to have a 'warm' lamp for use at night which gives them a feeling of domesticity, whereas they dislike the 'cool' lamp which they associate with an institutional character. These cultural differences cannot be dismissed but should be taken into consideration when designing the artificial lighting for tropical buildings. It is also a well-attested experimental fact that human beings with dark skin pigmentation also possess a greater degree of macular pigmentation in the eye, the result of which is to render them less sensitive to blue radiation. It is therefore possible that the colder, bluer light of the daylight type fluorescent lamp creates a different and possibly more pleasant visual sensation of colour to them than it does to the European.

Artificial lighting in tropical buildings has not up to the present developed a new vernacular. The 'package deal' type of lighting equipment characteristic of the general run of buildings in temperature civilizations has taken over in tropical buildings with little or no modification. At the moment, therefore, the techniques for lighting special buildings appear to be very much the same all over the world.

13
The Appraisal of Lighting

A systematic appraisal of the lighting of an interior is useful in a number of ways. It may be done by a group of experienced research workers in order to help to establish a new lighting standard or to develop a new design technique. The users of a building may be asked to appraise the lighting as a part of a study of their reaction to the building as a whole. Students of lighting design—architects or engineers—may be taken through a series of appraisals with the aim of heightening their awareness of the factors that contribute to a satisfactory lighting scheme, rather than of producing statistically significant results.

The appraisal of lighting demands a combination of careful physical measurements of luminance, illumination, colour, and other physical factors of the environment, together with subjective evaluations which must be conducted systematically and with as much attention to detail as possible. Physical measurements of lighting have been dealt with extensively in standard textbooks. Attention will therefore be concentrated more on the methods of making a subjective appraisal of lighting.

The essentials of a good lighting appraisal are that the subjective evaluations and the physical measurements should be made together and related both in the field and also subsequently by more careful analysis. Each factor of the visual environment requires separate evaluation, in some cases entirely subjectively, in other cases by a combination of subjective and objective evaluation, but rarely by physical evaluation alone.

ATTRIBUTES FOR APPRAISAL

The headings under which the appraisal of a lighting installation can be made are as follows:

(1) The adequacy of the quantity of light for the job to be undertaken. For example, if the lighting installation is in an office, judgement should be made of the adequacy of the light for reading, typing, operating punched card machines, or for whatever the critical visual task is judged to be. With experience, such judgements can be made without spending a long period in the room, but ideally enough time should be spent to be able to judge both the long-term as well as the short-term effects. Unfortunately this is often a counsel of perfection.

(2) The brightness distribution in the room, in particular as to whether the critical visual tasks have a preferential brightness, which is good, or whether there are distracting areas of high brightness which take the attention away from the work, which is bad; and also whether the brightness distribution gives an impression of pleasantness, gloominess, excessive stimulation, a soporific effect, etc.

(3) The relative amount of direct light from the windows or from the lighting fittings as compared with the diffused light reflected from all the surfaces of the room, and the extent to which modelling is influenced by these direct and indirect components of the lighting.

(4) Whether there is any suffusion of colour from highly coloured surfaces such as, for example, a red floor carpet producing a pleasant suffusion of warm light on the under surfaces of the human face, or a green carpet producing an unwelcome sickly effect.

(5) Whether there is any direct glare from windows or lighting fittings giving rise to discomfort.

(6) Annoyance, distraction, or a disabling effect due to

specular reflection of light sources in polished surfaces of a desk, office or industrial machinery, glossy wall surfaces or furniture surfaces, etc.

(7) The appearance of the windows, their shape in relation to the proportions of the room, their positioning, and the contrast, whether welcome or excessive, between visible sky and buildings outside on the one hand, and the interior room surfaces on the other.

(8) Appearance of the artificial lighting fittings, whether they are obtrusive by day and by night, whether they are properly integrated both as regards size, proportion, and positioning with the background of the room in general.

(9) The modelling of faces, of the texture of materials, and of the other relevant features of the work itself.

(10) The overall impression of the room and the character which it appears to have.

(11) The contribution of colour to this character of the room.

(12) The appreciation of scale and proportion in the room and the way in which the lighting or colour either modifies or appears to modify proportions which might otherwise be pleasant or unpleasant: whether walls or ceiling appear to advance or retreat because of light and colour, and whether this is desirable or not.

(13) In artificial lighting, whether there is any colour distortion of surface colours due to the quality of the lighting, and whether this distortion is welcome, hardly noticeable, or unwelcome.

OBSERVING IN A TEAM

The subject of evaluation of lighting is bedevilled by the fact that human beings are different and see lighting in a different way, and they appraise it differently depending upon their background culture, their long term and immediate past experiences, their general tolerance or intolerance of discomforts, and many other factors which result in individuals having different opinions.

Clearly if a lighting installation is intended for one person, and one person alone, then he is obviously the right person to make the appraisal provided that he is competent to do this.

On the other hand, if the room is to be used by a large number of people, say as an office, but exactly who they are is not known when the installation is designed, then a useful appraisal can be made if it can be demonstrated that the results of the appraisal will be in line with what the eventual occupiers of the office will themselves decide. Methods of statistical sampling are available which give some guidance, though only partial guidance, as to what should be done in such a situation. If we are considering the lighting of a hospital ward, for example, it would probably be statistically desirable to select a representative panel of nurses, doctors, and patients to make the appraisals, which might then be shown to be statistically related in some precise numerical way to the probable appraisals of the nurses, doctors, and patients who are going eventually to occupy the wards under consideration. Unfortunately such a statistical procedure is rarely practicable, and moreover it may not be satisfactory since nurses, doctors, and patients may not be very knowledgeable about lighting and may not know what things they should be looking for, even though their own comfort and amenity is under consideration.

A great deal has been written, and much more has been talked, about this question of who should make the lighting appraisals. Any individual can make his own appraisals; the appraisal at the end of this chapter was made by the actual occupants of a work-room. Where, however, the results of the appraisal are to be used for establishing that standards have been met, or for purposes of research, an organised team of observers is desirable. Experience has shown that appraisals are then best made by a small team of observers, comprising from 6 to 20 individuals, who have built up over a period a degree of skill and sensitivity. The position is analogous to the whisky or the tea industry. Blends of whisky or of tea are assessed by skilled panels of trained observers who know exactly what to seek and whose appraisals are shown by hard commercial results to be satisfactory.

The situation is exactly the same in lighting appraisals. It is important that the appraisers should have no vested interests in the results. They should be keen that the results be accurate and valid, and so they should be motivated in this way, but they should have no other motivation, no bias towards the manufacturers', the contractors' or the users' interests. Such ideal teams of trained observers are difficult to assemble and difficult to train. Nevertheless, useful appraisals can be made by any group of keen and informed observers, such as the potential users of a building, provided they are ready to learn and apply to the best of their ability the methods of appraisal.

A distinction should be made between appraisals for research or design purposes, and those for purely educational or informative reasons. Appraisals are a valuable method for assisting in the formulation of lighting standards, but these demand careful planning and evaluation in order that the results they give will carry sufficient conviction to ensure that the standard will be held in respect. Educational and informative appraisal also need careful planning if they are to be of use and value, but they do not require the same subservience to the demands of current statistical methodology.

UNDERTAKING AN EDUCATIONAL APPRAISAL

The group of observers should each be provided with a pro forma sheet which will pose questions about the relevant aspects of the lighting. The observers will each be asked to record individually a value judgement about each aspect in turn. This judgement can most reliably be made by choosing a point on a scale either:

(a) in words   (b) with numbers   (c) a linear scale

very good			⇧	4		good —
good	x		good			
				3	x	
fair				2		⇦
			bad			
bad			⇩	1		bad —

Some scales have two undesirable extremes, e.g. from 'much too bright' through 'just right' to 'much too dark' and although this must be recognised, it is generally advisable to avoid scales with an odd number of steps and a neutral centre point, so to prevent the observer from taking refuge in a non-committal reply. Observations can often usefully be supplemented by verbal comments.

Observers should assess independently and not discuss any of their findings until after the completion of the appraisal. The variance of the opinions of individuals of a team is important information. For example, there might be unanimity that a room is pleasant, but a wide variance about the adequacy of the lighting for its purpose. It is of importance to discover this variance and to find if possible the reasons for it.

First impressions should preferably be recorded in the observer's own words. Once these first impressions have been recorded, the observers should be given plenty of time to adapt themselves to the situation, and if a specific visual task, such as reading a book, reading a chalkboard, typing, etc. is involved, they should if possible be set to this task and given sufficient time to be thoroughly adapted to the conditions prevailing. They should then be asked to make their appraisals of the first five attributes listed above. In general these are matters which observers are able to assess fairly readily, particularly after they have had some experience.

The next six attributes are much more difficult. Questions of proportion, obtrusiveness, modelling, etc. require a great deal more experience before any confidence can be built up, but this confidence does eventually come and even if the team are completely inexperienced, they should be asked to think hard and make a judgement even if they may themselves have little confidence in it. The choice of criteria will have to be made with considerable care. It is not possible within the scope of this very brief treatment to give more than an indication of the solution to the difficulties involved. For example, the criteria selected for assessing the 'character' of a hospital ward would be quite different from those selected for a school classroom. In the case of a hospital ward the observers would be looking for qualities of the lighting which provided either excessive stimulation on the one hand or excessively soporific effect on the other. On the other hand, in a school classroom, while the polarities of stimulation and sedation may also be relevant, it is important in addition to assess whether the lighting is distracting on the one hand or whether it is designed to direct attention to where it is required, the chalkboard or the desk, or the model in an art room, for example, and so the assessment of the 'character' of the room would have to introduce attributes of distraction on the one hand and of concentration on the other.

EXAMPLE OF AN APPRAISAL OF DAYLIGHTING USING A
LINEAR SCALE METHOD OF APPRAISAL

Subject. A general workroom used at different times for writing, reading, typing, drawing, sewing by hand, and by machine.

Lighting. By two small windows in opposite sides, high reflectance of interior surfaces, giving a minimum daylight factor of 0·7% and an average over the working area of 1·5%. Ceiling reflectance 80%, upper walls 80%, three lower walls 30%, one wall (upper and lower) grey 50%, floor 16%.

Ceiling and three upper walls white, three lower walls Munsell

5Y. 6/4, one wall N. 7·5 patterned, floor 5R. 4·5/2 (textured carpet).

Luminance of work (cloth on sewing machine), table, and floor, 10, 5, and 2 ft-L respectively.

Sky luminance 1300 ft-L.

Luminance of wall in view from work table (grey wall) 15 ft-L.

1. Your first impressions of the room.

| very unpleasant | very pleasant |

not enough light for the job | just

too much light | right

room too dark | just

room too bright | right

2. Sit down taking a little time to adapt to each job in turn;
For reading, is the lighting

totally inadequate | completely adequate

writing,

totally inadequate | completely adequate

typing,
hand sewing,
machine sewing,

3. From where you are now at the sewing machine is the general brightness in the room

very unpleasant | very pleasant

very gloomy | very cheerful

sleepy | stimulating

cold | warm

Is the brightness of the work in relation to the rest of the room

too dark

| just right

too bright

Is there anything in the room which is very bright and is this

very distracting | not distracting

Is there anything in the room which is very dark and is this

very distracting | not distracting

Can you comment on the above? | Window in view, sky too bright.

4. Shade the work from the win-   none   all
   dows with your hand and esti-
   mate how much of the light on
   the work comes directly from
   the window

5. Is any colour reflected off the   none   a lot
   floor, ceiling or walls or furni-
   ture on to the work; is this
   pleasant   unpleasant   pleasant
   on to people's faces   none   a lot
                                      unpleasant   pleasant
   on to something else, e.g.   none   a lot
   ceiling   unpleasant   pleasant
   Can you comment?   No noticeable reflected light except green
                                      on ceiling near the window from sun on
                                      grass outside

6. Is there glare from the windows   intolerable   no glare
   from anywhere else   intolerable   no glare
   Can you comment?   One window shielded by wall, the other
                                      visible and too bright

7. Are there reflections on the   many   none
   table top or in your machine   very   not at all
   and are they   annoying   annoying
                                      very   not at all
                                      disturbing   disturbing

8. Are the windows well propor-   not well   very well
   tioned in relation to the room   proportioned   proportioned
   well positioned from your point   not well   very well
   of view   positioned   positioned
   Is there brightness contrast be-   great   no
   tween the room and what you   contrast   contrast
   see through the window
   Is this good or bad   very bad   very good
   Is the view through the window   inadequate   adequate
                                      unpleasant   pleasant
                                      (1) too bright, sky only
                                      (2) no view; the view is pleasant when seen

9. Is the appearance and model-   very much
   ling of faces   too harsh   just
                                      very much   right
                                      too soft
   Is the texture of material shown   very poorly   very well
   up

10. Can you describe the overall character of the room in a few words

Soft lighting, very pleasant marred only by bright sky through high small window

Is this character influenced
by—the lighting   7
    —the colouring of the walls   2
    —the furniture   1
(try to assess the proportion due to each, out of 10)

11. Does any wall or the ceiling give you the impression that it recedes or approaches because of the lighting or the colour?

Yes, wall opposite me advances

Has it changed as a result of a lighting or colour change which you remember?

Yes, redecoration

What was the change; was it good or bad?

Light grey replaced dark red which receded. Pleasant, room now better proportioned

12. Please don't look back at the first question.

Is your last impression of the room that it is

very unpleasant	very pleasant
not enough light on the work	
too much light on the work	just right
room too dark	
room too bright	just right

Have you any further comments on the lighting or colouring or any other visual feature of the room?

Placing of work table too far from windows, would be better to have more direct light on sewing machine and also to have a view through the lower window while shielding the upper, which causes some sky glare

COMMENTS

Although the daylight factor in the working area of the room is low (1·5% average, 0·7% minimum) as compared with the standard required in a comparable room in a school (2% minimum), the initial impression is that the lighting is just a little below adequate, although the final impression after the appraisal is completed rates it a little lower than at first.

The detailed appraisal rates the amount of light adequate for relatively easy tasks like reading and writing but less than adequate for sewing. Since the illumination at the time must have been of the order of 10–15 lm/ft², it is surprising that an even lower rating was not made. This may well be due to the overall impression of brightness, e.g. on the walls, having influenced the appraisal of the light on the work.

The room rates high on brightness appearance, the only bad feature being a high window through which the sky is visible, and this causes some distraction and glare. Reflections are no problem, presumably because the only primary source which could cause these, the high window, is not in a position to cause reflected glare.

The window appraisal has called for a distinction to be made between the two windows, one gives glare and no pleasant view, the other is shielded and causes no glare but the pleasant view outside is also lost as a result.

Modelling and revealing of texture is on the soft side and the comment is made that it could be sharpened up by positioning the work table to get more direct light from the lower window.

The lighting is seen to govern the character of the room, which is pleasant but apparently rather neutral, neither soporific nor stimulating.

A wall coloured light grey in a redecoration is now considered to 'advance' whereas previously it 'receded' when it was dark red. This is considered an improvement as the room now appears better proportioned.

The available physical measurements do not permit a great

deal of additional comment. The luminance ratios of task: immediate surround: general surround (10 : 5 : 2) as seen when
sewing are close enough to 10 : 3 : 1 for conformity; apparently
these features are satisfactory. The use of white walls minimises
sky glare but this is still a matter for unfavourable comment
with a fairly bright 13,000 asb (1300 ft-L) sky. The grey wall
seen from the work table has a slightly higher luminance 150 asb
(15 ft-L) than the work, but it is not specially commented on
(Q. 3).

The colour treatment is in the red/yellow/brown range and is
rated 'warm'. The questionnaire, however, is not sufficiently well
drafted to permit any indication as to whether this is liked or
not, but the overall rating of 'pleasant' suggests that it is. The
colour treatment, however, is given only a 2/10 influence on the
character of the room as compared with the lighting.

THE CORRELATION OF SUBJECTIVE AND OBJECTIVE ASSESSMENTS

The end point of any lighting appraisal is to determine to what
extent the lighting is suitable for its purpose and to what extent
it fails in one way or another. Whatever the readings of physical
instruments may be, it is the human appraisal which is the final
arbiter of the success of the installation. Consequently any physical measurements which may be made must take their proper
place in the priorities and must not be used as the basis for any
final decision on the satisfactoriness of the installation.

Nevertheless in some cases there are standards or regulations
which specify in quantitative terms what the lighting should be.
In such cases any appraisal must include physical measurements
to a satisfactory standard of accuracy. For example, school
building regulations in Great Britain (at present, 1968) require
that in any teaching area the available illumination shall not be
less than 10 lm/ft^2 at any time. Clearly any appraisal of the
lighting installation must include a measurement of the illumination at strategic points.

Subsequently these measurements can be correlated with the subjective assessments of adequacy. It may then be found, for example, that the lighting was judged inadequate even though the statutory standard had been satisfied. This would suggest looking for adaptation effects or might well raise in the mind the adequacy of a standard which deals only with quantitative matters.

For such a comparison to be of value, however, the physical measurements would have to be accurate, that is, their accuracy would have to be known before it would be wise to question the validity of the standard rather than the reliability of the measurements.

A useful adjunct to an appraisal is the taking of photographs and the preparation of sketches as a record of what was seen. Reference back will often resolve inconsistencies, showing some factor which was missed during the appraisal. Photographs are not necessarily accurate. At best a paper print can only record a very limited range of brightness—a transparency can do better—but neither can record constancy effects and only to a limited extent the effects of local adaptation and simultaneous contrast of neighbouring areas. A good sketch can record the true visual effect if the artist is well trained.

The sorting and analysis of the data of an appraisal can be undertaken at various levels—by a statistical analysis with all the paraphernalia of punched cards and computers—or more simply. The statistical study will be more rigorous and may be more revealing but a simple study of the data may show quite quickly what are the most interesting features of the appraisal.

When there are data from a group of appraisals, useful and interesting comparisons can be made. Among the points which can be looked for in the comparisons are the following:

(1) What governs the judgement of the adequacy of the lighting for the work—is it the amount of light on a horizontal surface at the work, or on a vertical surface, or do other factors come into account?

(2) What governs the impression of quantity of light in the room—is it the light on the working area, or the light on the walls and surroundings generally?

(3) How does a view of the light sources, whether windows or lamps, affect the judgement of the amount of available light —do small bright sources like filament lamps give an impression of more light about, or of less?

(4) If the lighting is concealed and integrated with the structure, does the result look as though there is more light about than when the fittings or windows are visible—would a designer who integrated his lighting need to convince a client that there really was sufficient light, or would it be self-evident?

(5) Does a little glare give a favourable or an unfavourable first impression to be subsequently modified after adaptation and experience—and if so, how much glare has a marked result? (This effect may be important because first impressions often are of over-riding importance—should the architect be judged at the official opening of the building or by the occupants at a later stage?)

(6) To what extent does colour affect both the initial and the eventual impression—does strong colour excite a favourable first impression, to be replaced by annoyance or distraction, or does it 'grow on one' like an acquired taste?

All these questions will, of course have to be answered first by 'it all depends' upon particularities, but they will serve to illustrate the educational and informative value of appraisals, particularly in the qualitative aspects of lighting design.

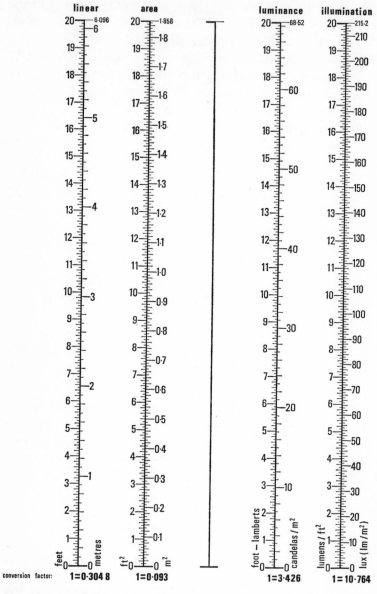

linear	area		luminance	illumination
20 — 6·096	20 — 1·858		20 — 68·52	20 — 215·2

Appendix. Metric Conversion Scale. (Reproduced by permission of the Polytechnic, Regent Street, London.)

Short Bibliography

This bibliography is a short list of additional reading on the main subjects dealt with by each chapter in turn. Most of the recommended books include further references in their own subjects. Books are published in London unless otherwise stated.

CHAPTER 1
Art and Illusion (especially Part Three), E. H. Gombrich (Phaidon, 1962).
Lighting and Design, Leslie Larson (Whitney, New York, 1964).

CHAPTER 2
International Lighting Vocabulary, Commission Internationale de L'Eclairage (Paris, 1967).

CHAPTER 3
Art and Visual Perception, Rudolf Arnheim (Faber, 1967).
Sight, Light and Work, H. C. Weston (Lewis, 1962).
Architectural Physics—Lighting, R. G. Hopkinson (H.M.S.O., 1963).
Eye and Brain, R. L. Gregory (Weidenfeld and Nicolson, 1966).

CHAPTER 4
Daylighting (British Standard Code of Practice CP3: Chapter 1: Part 1—1964), British Standards Institution.
Daylighting, R. G. Hopkinson, P. Petherbridge, J. Longmore (Heinemann, 1966).
Sunlight in Buildings, R. G. Hopkinson (Ed.) (Baucentrum International, Rotterdam, 1967).

CHAPTER 5

Lamps and Lighting, H. Hewitt and A. S. Vause (Eds.) (Edward Arnold, 1966).
Interior Lighting Design Handbook, British Lighting Council (2nd Edn., 1966).
The Ergonomics of Lighting, R. G. Hopkinson and J. B. Collins (Macdonald, 1969).

CHAPTER 6

Integrated Daylight and Artificial Light in Buildings, Building Research Station Digest No. 76 Second Series (H.M.S.O., 1966).
Lighting during Daylight Hours, Illuminating Engineering Society Technical Report No. 4 (1962).

CHAPTER 7

Colour in School Buildings, Department of Education and Science Building Bulletin No. 9 (4th Edn.) (H.M.S.O., 1968).
Lighting in Schools, Department of Education and Science Building Bulletin No. 33 (H.M.S.O., 1967).

CHAPTER 8

Hospital Lighting, R. G. Hopkinson (Ed.) (Heinemann, 1964).
Studies in the Function and Design of Hospitals, Nuffield Provincial Hospitals Trust (Oxford University Press, 1955).
Hospital Lighting, Technical Report No. 12. Illuminating Engineering Society, London, 1968.

CHAPTER 9

The Lighting of Office Buildings (Post War Building Studies No. 30), H.M.S.O. (1952).
Better Office Lighting (The Electricity Council, 1966).

CHAPTER 10

The Lighting of Factories (Factory Building Study No. 2), Building Research Station (H.M.S.O., 1959).
The Colouring of Factories (Factory Building Study No. 8) Building Research Station (H.M.S.O., 1961).

CHAPTER 11
Lighting for Life, J. Boud (Builder).

CHAPTER 12
Daylighting, Chapters 20 and 21 (and the further Bibliography in this book), R. G. Hopkinson, P. Petherbridge, J. Longmore.

GENERAL READING
Recommendations for Good Interior Lighting (The I.E.S. Code), Illuminating Engineering Society (1968).
Lighting in Architectural Design, Derek Phillips (McGraw-Hill, New York, 1964).

Index

adaptation, 46–51; colour, 50, 62
Ahrends, Burton, and Koralek, *Pls 43, 63*
Anderson, Charles, *Pl 50*
apostilb(asb), definition of, 39
apparent brightness, 40, 65–6, 151, *Fig 5*
Architects' Co-Partnership, *Pls 35–67*
Arnheim, Rudolf, 311
art galleries, *Pls 39–43*
artificial lighting, as supplement to daylight, *see* P.S.A.L.I.; calculations, 136–42; choice of source for, 133; distribution of, 136; glare from, 121, 143–6; in homes, 269–80; in hospital wards, 204–7; in factories, 251–6; in schools, 183–90; in the past, 33–34; in the tropics, 294; role of, 122
attention, attracted by brightness and colour, 49, 55, 56, 175, *Fig 7*

bathrooms, 269
bedrooms, 265, 278
Bickerdike, John, *Pl 40*
blinds, window, 28, 92–3, 99, 176, 183, *Pls 24–7*
Boud, J., 313
brightness, apparent, 40, 65–6, 151, *Fig 5*; constancy, 27, 65–6; as luminance, 39; distribution in sky, 81–5; of fluorescent fittings, *Pl 32*; variety in, 25

British Lighting Council, 312
B.R.S. Internally Reflected Component Tables, 114
B.R.S. Sky Component Protractors, 111–14, *Fig 12*
building lighting, 122
Burnet, Sir John, Tait and Partners, *Pl 56*
'bürolandschaft', 162, 235
BZ (British Zonal) classification, *Fig 15*

ceilings, luminous, 147
chalkboard, lighting of, 189, *Pls 45, 46*
children's vision, and lighting, 168
C.I.E. (Commission Internationale de l'Eclairage), standard overcast sky, 81; vocabulary, 311
clinical examination rooms, 213
coefficient of utilisation, 106, 137–8, 291
Collins, J. B., 312
colour, 25, 30–1; adaptation, 50, 62–4 ; appearance of colours under fluorescent lighting, 160; appearance of fluorescent lamps, 158; constancy, 66; enhanced by lighting, 52; in hospitals, 218; in offices, 227; Munsell system for, 30, 40, *Fig 2*; rendering, 31, 50, 62–4, 124, 156, 207, 236, *Fig 14*; temperature, 84–5, 191, *Fig 14*
comfort, thermal, 260–2; visual, 56–7